AIR POLLUTION

IN TWO PARTS

Part A—Analysis

JOE O. LEDBETTER

COLLEGE OF ENGINEERING
THE UNIVERSITY OF TEXAS
AUSTIN, TEXAS

MARCEL DEKKER, INC., New York   1972

MARCEL DEKKER, INC.

95 Madison Avenue, New York, New York 10016

LIBRARY OF CONGRESS CATALOG CARD NUMBER:
77-160112   .

ISBN: 0-8247-1405-9

Printed in the United States of America

To my lovely wife,

Nan Thompson Ledbetter, BA, MA, PhD

PREFACE

Engineering control of air pollution is carried out through the same procedural attack that engineers have historically followed; namely, the analysis of possible problems and the application of effective controls. These facets are covered in two parts, Part A: Analysis, and Part B: Prevention and Control, in order to offer books of manageable size and to permit persons involved in only one phase to obtain the materials which they need. College courses in Air Pollution Control often divide along these topic lines.

Analysis treats the measurement of the air pollutant and the evaluation of the problem presented by the pollutant. Prevention deals with the design, the installation and the effectiveness check of the preventive method--whether the method is a technique or equipment. The two volumes have developed from my teaching and consulting experiences over a ten-year period. There are twenty-four consecutively numbered chapters in the two books, ten in the first volume and fourteen in the second.

The understanding of any technical subject rests on a basic understanding of what the subject is about. The necessary background information and vocabulary are given in Chapter 1. The sources of air pollution and the amounts from these sources are summarized in the second chapter.

The third chapter treats atmospheric transport of pollutants. All of the major air pollution problems have come during periods of poor atmospheric mixing or as Donne said, "..; Aire that moveth not, poysoneth;.." The use of statistics is important in the mathematical treatment of transport and in air pollution sampling and sample analysis. Sufficient statistical descriptions for the purposes are covered in Chapter 4.

v

Effective analysis and prevention of air pollution are based
on the principal properties of the gaseous and particulate pollutants.
A brief review of this important subject is given in the fifth chapter. Chapter 6 is concerned with the actual sampling of air pollutants, including how, when, where, what, and how much.
Conventional sampling consists of removing the pollutant from a
known volume of air.

Chapter 7 describes sample analysis methodology for identifying and quantifying pollutants. The analytical principles are
given broad coverage in this important subject. The ninth chapter
presents the most used "cookbook" tests for the specific pollutants
generally considered. The importance of the physical properties
of particles, especially size, merits a separate chapter, the eighth.

The physical, chemical, and plant damage effects of air
pollution are much better documented than are the human health
effects, and this fact is apparent in Chapter 10. The ambient
levels of noise and pollen can be shown by toxicological tests to
cause physiological damage. The ozone and oxidant levels during
severe Los Angeles smog come within an order of magnitude of
toxic levels. Other health damage from air pollution is proved only
by the statistical methods of epidemiology.

Air pollution abatement is vital to our continued well-being.
Our control efforts will not be able to restore our entire atmospheric
environment to its pristine state within the foreseeable future, but
we do not have to live in nasty, noisy, unhealthful conditions.
Achieving the proper balance between economic development and
feasible control is perhaps the most difficult air pollution problem
today. The balancing decisions are necessarily political decisions,
but engineers have an important role in helping to inform the decision-
makers as well as in carrying out the control. My hope is that this
book will aid the engineer in his job.

My family, my students, and my colleagues have been long-
suffering during the production of this book and the least that I

should do is to make grateful acknowledgment of this fact.

Austin, Texas   Joe O. Ledbetter
February 1972

CONTENTS

AIR POLLUTION

PART A—ANALYSIS

Chapter 1

BACKGROUND FOR ANALYSIS

I. INTRODUCTION

Air pollution analysis is a necessary first step in the engineering control of air pollution. For economically planning and
carrying out the prevention of an air pollution problem, the pollutant must be well defined in nature and amount. This volume
is concerned with the terminology, methodology, and symptoma-
tology involved in the delineation of air pollution problems. The
information presented has been selected not for completeness,
which is impossible, but for expected usefulness in evaluating
air pollution situations. Where theoretical treatment is not deemed
requisite to the performance of an analysis, it is often omitted in
the interests of brevity.

II. DEFINITION OF AIR

A knowledge of the composition of air is required in any
study of air pollution. The composition of air varies somewhat
with local conditions; however, the average concentrations of
the various components can be rather narrowly defined. Table 1-1
shows the makeup of dry air. Of course air is not normally dry
but has a relative humidity of perhaps 50%. At 50% relative
humidity and $70^{\circ}F$, the percent (by volume) of moisture in the air
is about 1.15.

Table 1-2 lists some of the properties of air that are useful
in air pollution engineering. The apparent molecular weight is

1

TABLE 1-1

Composition of Normal Ambient Air (Dry)

Component	By volume	By weight	Total Mass (gm X 10^{-20})
N_2	78.084 ± 0.0004 %	75.51 %	33.648
O_2	20.946 ± 0.0002 %	23.15 %	11.841
A	0.934 ± 0.0010 %	1.28 %	0.6555
CO_2	0.033 ± 0.0010 %	0.046%	0.0233
Ne	18.180 ± 0.040 ppm	12.50 ppm	6.36×10^{-4}
He	5.240 ± 0.004 ppm	0.72 ppm	3.70×10^{-5}
Kr	1.140 ± 0.010 ppm	2.90 ppm	1.46×10^{-4}
Xe	0.087 ± 0.001 ppm	0.36 ppm	1.80×10^{-5}
N_2O	0.500 ± 0.110 ppm	1.50 ppm	7.70×10^{-5}
CH_4	2 ppm	1.20 ppm	6.20×10^{-5}
H_2	0.5 ppm	0.03 ppm	2.00×10^{-6}
O_3	0.01 ppm	--	--
Rn	10^{-13} ppm	--	--

the weighted mean obtained after summing the products of the fractions of various gases and their individual molecular weights. Note that moist air is lighter than dry air because the molecular weight of water is only 18. There is an impressive amount of air on earth, enough to exert more than 1 ton/ft^2 on all of the earth's surface, nearly 6×10^{15} tons.

A general description of air might be that it is a continuous, compressible, ideal fluid. Some purposes require or permit the modification of this definition; for example, small particles settle as if air is not continuous, and certain problems are simplified, yet solved satisfactorily, by assuming air to be incompressible.

The earth's atmosphere may be divided into two zones according to height above the earth, the troposphere (below about

TABLE 1-2

Properties of Air

Property	Value	Units/Comments
Mean-free-path, λ	6.63×10^{-8}	m $\quad \lambda = 1/(\pi\sqrt{2}\ N\ d^2)$ N = molecules/volume d = molecular diameter
Thermal conductivity	5.68×10^{-5}	cal/cm^2-sec-$^\circ$C/cm
Specific heat, c_p	0.24	cal/$^\circ$C-gm (or Btu/$^\circ$F-lb) Constant pressure
Specific heat, c_v	0.17	cal/$^\circ$C-gm (or Btu/$^\circ$F-lb) Constant volume
Universal gas constant, R	1.9869	cal/$^\circ$C-gm mole-atm
Mole volume STP	22.4146	liters (or 359.0 ft^3/lb mole)
Mole volume NTP (20°C)	24.056	liters (or 385.3 ft^3/lb mole)
Apparent molecular weight	29.095	
Density, ρ (dry, STP) 50% RH, NTP Liquid, -147°C	1.293 1.20 0.92	gm/liter gm/liter (0.75 lb/ft^3 NTP) gm/cm^3
Dynamic viscosity, μ	170.8	μp $\quad \Delta\mu \simeq 0.36\ \Delta^\circ C + 5$ $-\Delta\mu \simeq 0.6\ \Delta^\circ C + 5$
Kinematic viscosity, ν	13.2	centistokes (cs) $\nu = \mu/\rho \quad$ cs = 0.01 cm^2/sec
Relative humidity, RH	100 p_v/p_s	p_v = vapor pressure (vp) p_s = saturated vp (mm Hg) $\log_{10} p_s = 9.036 - 2284/T$ $\quad$ T in $^\circ$K
Velocity of sound, v_o STP	1087	fps $\quad v = v_o \sqrt{(^\circ K/273)}$
Refractive index, m	1.00029	Sodium light (524 nm)
Solar wavelength, λ_s	524	nm $\quad$ Characteristic

11 km) and the stratosphere (above the troposphere). The troposphere is that part of the atmosphere where, on the average, the
temperature decreases with height. The stratosphere has but
little change of temperature with height and no cumulo-nimbus

clouds. Air pollution analysis is ordinarily concerned with the
lower portion of the troposphere, up to about 8 km, or through
about 70% of the total air by weight.

III. FUNCTIONS OF AIR

The importance of air can hardly be overemphasized,
although that of clean air may well be. Air serves so many
functions that it would be practically impossible to cover all
of them; some of these functions are briefly described here.

Air serves as a storehouse and carrier for oxygen. It not
only brings oxygen to us but also carries away the waste products
of the oxidation reactions as well as other wastes injected into
the air. It is the abuse of the carrying (and diluting) function
that causes air pollution problems. The chemical reactions that
take place in our air space are a mixture of desirable and unde-
sirable ones. On the one hand, pollutants are broken down into
simpler, less obnoxious compounds, while on the other, reactions
form noxious compounds from inoffensive substances.

In addition to gases air also carries sound and light.
Without air no sound could exist, and while light penetrates air
it does so with a loss by scattering; visibility is limited by air
scatter to about 150 mi., and a shadow is not perfectly dark
because of scattered light.

The atmosphere absorbs most of the damaging radiations
that would otherwise reach earth from space, both corpuscular
(electrons, protons, and so on) and electromagnetic [ultraviolet
(UV), γ rays, and even highly energetic cosmic rays].

IV. DEFINITION OF AIR POLLUTION

Despite the risk of proliferation of the many definitions for
air pollution, a different definition is presented here in an attempt
to make the definition as simple as it should be. Air pollution is

the presence in the air of any abnormal material or property that reduces the usefulness of the air resource. Although this definition does not expressly state ambient or outdoor air, because the definition should not be so limited, it is understood to refer to outdoor air unless specifically stated otherwise.

. The Engineers' Joint Council has arrived at a rather verbose definition: "Air pollution means the presence in the outdoor atmosphere of one or more contaminants, such as dust, fumes, gas, mist, odor, smoke, or vapor, in quantities, of characteristics, and or duration such as to be injurious to human, plant or animal life or to property, or which unreasonably interfere with the comfortable enjoyment of life and property."

There are many legal definitions for air pollution, nearly all of which contain words such as thereby, thereof, whereas, and so on, words that engineers do not normally include in a working definition.

V. CLASSIFICATION OF POLLUTANTS

A list of air pollutants should cover not only the classic items smoke, dusts, smog, and other particulates, sulfur dioxide, oxides of nitrogen, oxidants, and other gases, but also other items that satisfy the definition, such as noise, heat or cold, fog, radioactivity, excess radiation (UV, visible, and microwave), and similar properties. Air that becomes less useful through the removal of a normal constituent is also polluted.

Pollutants should be amenable to some classification for ease of reference. The classifications used here, which are actually lists of definitions, are based upon the physical properties of the pollutants and the history of their formation or production. A definition of fumes that excludes the mists from fuming sulfuric acid may seem strange, but this is the definition of fumes commonly used by engineers involved in air pollution control.

Naturally, no definitions could be set down that would encompass all the usages of a word. Some will incorrectly apply the term fumes to acid mists and even to the gases and vapors of auto exhausts.

A. American Standards Association Definitions

The ASA defined the common forms of air contaminants in 1936. These definitions, which are probably as good as any, are:

Dusts: Solid particles generated by handling, crushing, grinding, rapid impact, detonation and decrepitation of organic or inorganic materials such as rock, ore, metal, coal, wood, grain, etc. Dusts do not tend to flocculate except under electrostatic forces; they do not diffuse in air but settle under the influence of gravity.

Fumes: Solid particles generated by condensation from the gaseous state, generally after volatilization from molten metals, etc., and often accompanied by a chemical reaction such as oxidation. Fumes flocculate and sometimes coalesce.

Mists: Suspended liquid droplets generated by condensation from the gaseous to the liquid state or by breaking up a liquid into a dispersed state, such as by splashing, foaming, and atomizing.

Gases: Normally formless fluids which occupy the space of enclosure and which can be changed to the liquid or solid state only by the combined effect of increased pressure and decreased temperature. Gases diffuse.

Vapors: The gaseous form of substances which are normally in the solid or liquid state and which can be changed to these states either by increasing the pressure or decreasing the temperature alone. Vapors diffuse.

B. Other Definitions

Several terms other than those defined above are needed to characterize air pollutants. Some of these in common usage are given here:

Fly ash: Particles of ash which become entrained in the combustion gases and are carried into the air

Aerosols: An air or gas suspension of particles (usually less than 50 μm in diameter)

Fog: An aerosol of liquid droplets near the ground as distinct from clouds

Smoke: Carbon or soot particles, often less than 0.1 μm in diameter, that result from the incomplete combustion of carbonaceous materials such as coal, oil, tar, and tobacco

Smog: Mixture of smoke and fog as coined in England. Presently used to describe photochemical aerosols such as those in Los Angeles

Haze: Suspension of small (fractional micrometer) particles in the air which makes distant, large objects indistinct

Smaze: Mixture of smoke and haze. Suggested for photochemical smog but not widely used

Radioactivity: The nuclear disintegrations atoms undergo in giving off ionizing radiations

Radiation: Electromagnetic radiations plus those corpuscular emissions classed as "rays"

Noise: Unwanted sound. Sound is the effect that rapid, local fluctuations of the atmospheric pressure have on the ear.

VI. UNITS FOR QUANTIFICATION OF AIR POLLUTION

Stack gas concentrations of particulates, which are usually rather high, are given in grains per cubic foot (grpcf or gr/cf) or milligrams per cubic meter (mg/m^3), and those of gaseous materials in percent by volume ($\%_v$) or parts per million by volume (ppm_v). Occasionally, other units are used; for example, gaseous concentrations are in weight units and stack concentrations in pounds per 1000 lb of flue gas at a given percent of excess air. The cubic feet or cubic meter applies at stack conditions of temperature and pressure unless otherwise stated. If the concentrations are listed at standard conditions ($0^\circ C$ and 1 atm), the abbreviations become scf and scm, or at normal conditions ($20^\circ C$ and 1 atm), ncf. There is common usage of the normal conditions for the standard conditions (scf at $20^\circ C$). This practice should have been avoided because of the long-standing definition of STP. Of course, parts per million will not change as the volumes change since the ratio is unaffected.

Ambient air concentrations are usually much more dilute than those for stack gas and are reported in micrograms per cubic meter ($\mu g/m^3$) for particulates, and frequently for gaseous materials as well; however, gaseous concentrations are more likely to be in parts per million by volume (ppm_V). Sometimes ambient concentrations are given in other units in order to obtain small whole numbers; namely, milligrams (mg, 10^{-3}gm), nanograms (ng, 10^{-9} gm), and picograms (pg, 10^{-12} gm) per cubic meter or parts per hundred million ($pphm_V$) or parts per billion (ppb_V).

EXAMPLE 1-1: Convert 40 $\mu g/m^3$ of sulfur dioxide at NTP to parts per million.

Mole volume NTP = 24.06 liters (from Table 1-2)
Molecular weight of sulfur dioxide = 64

$$40 \frac{\mu g}{m^3} \times \frac{1}{64 \times 10^6} \frac{gm\ mole}{\mu g} \times 0.024 \frac{m^3}{gm\ mole} \times 10^6 \frac{parts}{million\ parts}$$

$$= 0.015\ ppm_V = 1.5\ pphm_V$$

Special units and indices for measuring air quality are defined in Chapter 7, including those for noise, dustfall, dirtiness, visibility, and so on.

VII. EXTENT OF AIR POLLUTION

The extent of air pollution today depends upon viewpoint, including the definition of what constitutes air pollution. Practically everyone would change some aspect of the air around him. Some air pollution problems are spread over a wide area; for example, the fallout from nuclear detonations in the atmosphere is worldwide. On the microscale a person with body odor may pollute only a few cubic feet of the air around him.

A. General Problems

It is doubtful that there are any real air pollution problems that threaten health worldwide; probably only psychological and political problems exist on such a scale. The pollutants that

could possibly fit this category are fallout, DDT, and carbon
dioxide. Certainly, these have all been cited by various observers
as the agents that will cause our downfall.

Fallout of man-made radioactive materials from nuclear
tests dates from July 1945 when the first atomic bomb was deto-
nated at Alamagordo, New Mexico. Although the amount of
radioactive fission products formed in an explosion is immense
(about 10^{11} Ci for a 20-kton burst), the activity decays rapidly
and dosages received by a population at a distance from the test
area are minimal, less than that from natural background radiation
and medical exposure. Experiments that have shown life-
lengthening to result from moderate amounts of radiation, and a
hundred times as many probable mutations from wearing trousers
rather than kilts as from all the fallout of the weapons tests, are
seldom cited but do refute the wild extrapolations made by several
doom prophets and widely quoted in the news media. The most
knowledgeable radiological health investigators have now focused
their attention on radiation from medical exposures and from
electronic appliances as the significant exposures to the population.

Pesticides in general, and DDT in particular, have been
widely castigated for their harmful effects. The principal offense
of DDT seems to be that it is found in all human and animal body
tissues used in pathology. Our detection systems for DDT are
extremely sensitive, as they are for fallout radiation. Public
Health Service research conducted with human volunteers in an
Alabama prison who ingested rather high concentrations of DDT
in their normal food and liquid intakes without showing detectable
effects should help allay our fears. There are some harmful
effects of pesticides--honey bees are killed; other helpful insects
are killed; and some birds have died or failed to reproduce because of pesticides, but many millions of human lives have been
saved from malaria alone by use of DDT.

Carbon dioxide may have increased about 10% in the past
50 years according to available data. Rather careful measurements
showed a 1.36% increase from 1958 to 1963. It has been theo-
rized that increased carbon dioxide will have a greenhouse effect
because of its affinity for infrared (IR) absorption and reradiation.
This warming could melt the polar ice caps and cause the oceans
to rise hundreds of feet in the next few thousand years. Other
observers do not adhere to this theory concerning an upset in the
carbon dioxide balance of the environment by the burning of fossil
fuels. Cadle and Magill point to the large amounts of carbon
dioxide dissolved in the waters of the earth and in limestone and
dolomite as reasons for the prevention of major long-term changes
in the atmospheric carbon dioxide. The increase in carbon dioxide could well cause increased photosynthetic activity which
would offset the trend; fossil fuels may be replaced as our energy
source; or climatic feedback in terms of rainfall changes and such
will more than likely balance the effect.

There is also concern by some observers that the reduction
in insolation by particulate pollution will lower the temperature
and cause another ice age. Perhaps the middle road will be
followed and the two effects will offset each other.

The major regional problems are climatic in nature. That
is, a climate may be too cold or hot, wet or dry, windy or calm,
or changeable for maximum use of the air resource. When such
areas are fully developed, the environment will be packaged or
controlled. The climatic phase of air pollution is not discussed
at any length here; the subject is left to meteorologists and books
on meteorology. Weather effects are quite costly in lost man
hours among outdoor craftsmen and to refugees from hurricanes
occurring in areas with normally good climates.

Thunderstorms, which are local or intermittent in nature
but general in scope, produce ozone and oxides of nitrogen along

with loud noises and fear among people to remain out of doors.
They limit the use of the air resource more than they pollute by
real agents.

Particulate problems are generally found in most locations
at one time or another; even the seashore has its salt spray. In
areas of limited vegetation, farm areas, deserts, and seacoasts,
for example, winds pick up dust from the ground and carry it in
amounts that cause problems.

B. Urban Problems

The most serious air pollution problems occur in urban areas
because there are more people in these areas--more people to
generate air pollution and more people to be exposed to the air
pollution generated. About 53% of the population of the United
States lives on about 0.75% of the land area. The population
density increases with the size of the city, as does the dirtiness
of the air (see Fig. 1-1). The National Center for Air Pollution
Control listed 65 cities of the United States according to their
air pollution problems (see Table 1-3). There should be an
additional column for meteorological factors, particularly the
frequency of stagnating conditions and perhaps for other parameters as well.

Other air pollutants follow the same trend noted above for
particulates, for example, oxides of sulfur (sulfur dioxide in
particular), oxides of nitrogen, benzene-soluble particulates,
carbon monoxide, aldehydes, and hydrocarbons. Most of these
problems arise from auto exhaust emissions as detailed in
Chapter 20. Figure 1-2 shows the concentrations of major gaseous
pollutants sampled by the Continuous Air Monitoring Program
(CAMP) in seven large cities. Lead has received considerable
attention in the last few years because of a controversy over the
seriousness of lead air pollution as a threat to health and the
desire to remove lead from gas in preparation for catalytic muf-

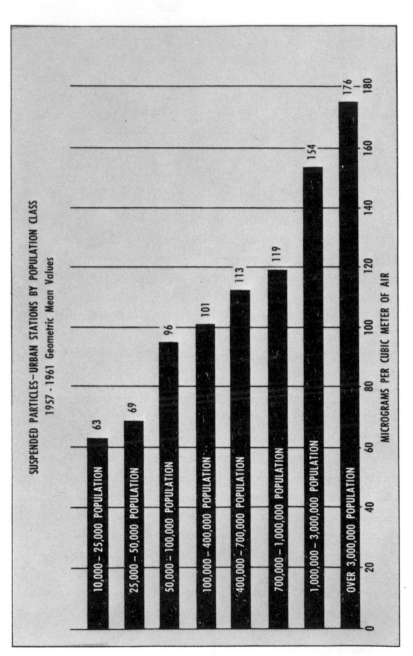

FIG. 1-1. Atmospheric dirtiness versus population. 1957-1961 geometric mean values. From "Troubled Air," U.S. P.H.S. Publ. No. 977, D.H.E.W., Washington, D.C., 1962.

flers. Despite the large amounts of lead that have been used in motor fuels, the principal sources of lead intake have always been food and beverages that are normally ingested. The urban levels of lead are usually well below 5 ug/m^3 for reported averages and that level gives only 0.05 mg/day (assuming 50% deposition, which is highly conservative), or about 15% as much as in food and beverages; however, concentrations found near freeways could conceivably double the total intake for some people and be of concern.

Although the exact causative agents have not been pinpointed, air pollution is probably responsible for much of the difference in incidence of respiratory diseases in urban and rural areas--lung cancer is about twice as frequent in cities as in rural areas. The benzene-soluble fraction is believed to be connected with lung cancer formation. Table 1-4 shows a summary of the airborne materials sampled during 1964-1965.

As viewed by the author, the most severe urban air pollution problems in order of severity are: (1) noise, (2) photochemical smog, (3) pollen, (4) carbon monoxide, (5) sulfur dioxide, and (6) dirtiness. Noise is placed highest on the list because society is either unable or unwilling to cope with this problem. Noise per se causes hypertension, deafness, and psychological effects, and sonic booms cause structural damage and irritate people subjected to them. Photochemical smog is believed to be the second worst offender because of its effects on the eyes and lungs, on visibility, and on materials in those cities where stagnation prevails long enough to permit buildups of ambient concentrations. This is really not a specific material but rather includes several, including nitrogen dioxide, oxidants, peroxyacetyl nitrate (PAN), polycyclic hydrocarbons, and so on. Pollen causes allergy problems in perhaps 10% of the total population. Carbon monoxide poisoning was described in detail more than 200 years ago, although not by name. Current research indicates

TABLE 1-3

NCAPC List of Urban Air Pollution Severity[a,b]

SMSA	Suspended particulates			Gasoline		Sulfur dioxide			Rank sum
	Average total	Geometric standard deviation	Average benzene-soluble	Total consumption	Auto emissions density	Arithmetic average concentration	Total emissions	Density of emissions	
1. New York	65	$20\frac{1}{2}$	62	64	63	65	63	55	$457\frac{1}{2}$
2. Chicago	$58\frac{1}{2}$	$14\frac{1}{2}$	50	63	49	64	65	58	422
3. Philadelphia	63	$7\frac{1}{2}$	$53\frac{1}{2}$	61	40	63	64	$52\frac{1}{2}$	$404\frac{1}{2}$
4. Los Angeles–Long Beach	53	$52\frac{1}{2}$	65	65	54	36	47	21	$393\frac{1}{2}$
5. Cleveland	$43\frac{1}{2}$	25	$32\frac{1}{2}$	55	60	$51\frac{1}{2}$	59	64	$390\frac{1}{2}$
6. Pittsburgh	62	$42\frac{1}{2}$	$53\frac{1}{2}$	57	25	41	61	48	390
7. Boston	45	$9\frac{1}{2}$	$57\frac{1}{2}$	59	59	54	$52\frac{1}{2}$	$52\frac{1}{2}$	389
8. Newark	33	$42\frac{1}{2}$	47	51	57	62	38	46	$376\frac{1}{2}$
9. Detroit	52	$48\frac{1}{2}$	35	62	55	$15\frac{1}{2}$	57	45	370
10. St. Louis	47	$9\frac{1}{2}$	48	58	$34\frac{1}{2}$	61	62	49	369

11. Gary-Hammond-East Chicago	56	$52\frac{1}{2}$	42	28	28	42	60	60	$368\frac{1}{2}$
12. Akron	$41\frac{1}{2}$	$58\frac{1}{2}$	$32\frac{1}{2}$	$25\frac{1}{2}$	53	$44\frac{1}{2}$	51	61	$367\frac{1}{2}$
13. Baltimore	$50\frac{1}{2}$	$20\frac{1}{2}$	56	49	$34\frac{1}{2}$	49	$52\frac{1}{2}$	43	355
14. Indianapolis	$60\frac{1}{2}$	5	61	40	58	$28\frac{1}{2}$	42	56	351
15. Wilmington	$58\frac{1}{2}$	$20\frac{1}{2}$	51	$20\frac{1}{2}$	$29\frac{1}{2}$	50	55	57	342
16. Louisville	57	$7\frac{1}{2}$	45	38	41	$32\frac{1}{2}$	58	59	338
17. Jersey City	46	$2\frac{1}{2}$	49	9	65	58	39	65	$333\frac{1}{2}$
18. Washington	$22\frac{1}{2}$	29	43	$52\frac{1}{2}$	46	$59\frac{1}{2}$	43	32	$327\frac{1}{2}$
19. Cincinnati	$41\frac{1}{2}$	13	38	42	48	47	45	51	$325\frac{1}{2}$
20. Milwaukee	$43\frac{1}{2}$	$48\frac{1}{2}$	21	45	51	$15\frac{1}{2}$	$35\frac{1}{2}$	$41\frac{1}{2}$	$301\frac{1}{2}$
21. Paterson-Clifton-Passaic[C]	17	$11\frac{1}{2}$	13	47	62	$59\frac{1}{2}$	40	54	304
22. Canton[C]	$60\frac{1}{2}$	33	60	8	23	$44\frac{1}{2}$	29	44	302
23. Youngstown-Warren	$54\frac{1}{2}$	38	52	14	16	39	41	40	$294\frac{1}{2}$
24. Toledo	$25\frac{1}{2}$	25	4	$25\frac{1}{2}$	56	38	50	63	287
25. Kansas City	48	$52\frac{1}{2}$	40	46	31	3	37	28	$285\frac{1}{2}$
26. Dayton	$35\frac{1}{2}$	$58\frac{1}{2}$	27	31	21	24	44	39	280
27. Denver	$54\frac{1}{2}$	56	$57\frac{1}{2}$	41	7	31	21	12	280
28. Bridgeport	$12\frac{1}{2}$	$2\frac{1}{2}$	$18\frac{1}{2}$	23	64	48	$30\frac{1}{2}$	62	280
29. Providence-Pawtucket	28	35	$24\frac{1}{2}$	27	42	57	22	$25\frac{1}{2}$	261
30. Buffalo	37	38	6	44	32	23	46	34	260
31. Birmingham	$50\frac{1}{2}$	63	55	32	26	9	11	13	$259\frac{1}{2}$

SMSA	Suspended particulates			Gasoline		Sulfur dioxide			Rank sum
	Average total	Geometric standard deviation	Average benzene-soluble	Total consumption	Auto emissions density	Arithmetic average concentration	Total emissions	Density of emissions	
32. Minneapolis–St. Paul	$9\frac{1}{2}$	38	10	$52\frac{1}{2}$	37	18	54	38	257
33. Hartford	$22\frac{1}{2}$	$45\frac{1}{2}$	17	33	50	40	$19\frac{1}{2}$	27	$254\frac{1}{2}$
34. Nashville	38	29	59	$20\frac{1}{2}$	44	21	$17\frac{1}{2}$	24	253
35. San Francisco–Oakland	6	57	28	60	36	13	$35\frac{1}{2}$	$17\frac{1}{2}$	253
36. Seattle	5	$42\frac{1}{2}$	$32\frac{1}{2}$	50	12	$25\frac{1}{2}$	56	29	$252\frac{1}{2}$
37. Lawrence–Haverhill^d									
38. New Haven	15	25	20	19	61	$44\frac{1}{2}$	26	$35\frac{1}{2}$	246
39. York	49	62	29	$5\frac{1}{2}$	10	$51\frac{1}{2}$	$19\frac{1}{2}$	$19\frac{1}{2}$	246
40. Springfield–Chicopee–Holyoke	4	$48\frac{1}{2}$	$15\frac{1}{2}$	11	43	35	34	50	241
41. Allentown–Bethlehem–Easton	34	17	14	10	13	55	49	47	239
42. Worcester	$9\frac{1}{2}$	$52\frac{1}{2}$	30	16	47	$44\frac{1}{2}$	13	22	$234\frac{1}{2}$
43. Houston	20	17	$11\frac{1}{2}$	56	45	3	48	33	$233\frac{1}{2}$
44. Chattanooga	64	31	64	7	11	22	$17\frac{1}{2}$	16	$232\frac{1}{2}$
45. Memphis	31	25	23	29	38	12	$32\frac{1}{2}$	$41\frac{1}{2}$	232

46. Columbus, Ohio	32	$20\frac{1}{2}$	22	37	52	7	24	37	$231\frac{1}{2}$
47. Richmond	17	$48\frac{1}{2}$	$32\frac{1}{2}$	15	27	37	23	$30\frac{1}{2}$	$230\frac{1}{2}$
48. San Jose	$22\frac{1}{2}$	60	63	34	22	8	$3\frac{1}{2}$	$4\frac{1}{2}$	$217\frac{1}{2}$
49. Portland, Oregon	$29\frac{1}{2}$	65	44	36	5	18	$6\frac{1}{2}$	$6\frac{1}{2}$	$210\frac{1}{2}$
50. Syracuse	39	33	41	17	$3\frac{1}{2}$	$28\frac{1}{2}$	$27\frac{1}{2}$	$19\frac{1}{2}$	209
51. Atlanta	17	17	26	48	33	11	$30\frac{1}{2}$	$25\frac{1}{2}$	208
52. Grand Rapids	40	61	$18\frac{1}{2}$	$12\frac{1}{2}$	20	34	9	9	204
53. Rochester	27	$11\frac{1}{2}$	$7\frac{1}{2}$	24	39	$28\frac{1}{2}$	$27\frac{1}{2}$	$35\frac{1}{2}$	$200\frac{1}{2}$
54. Reading	$35\frac{1}{2}$	6	38	2	$3\frac{1}{2}$	56	25	$30\frac{1}{2}$	$196\frac{1}{2}$
55. Albany-Schenectady-Troy	7	55	$7\frac{1}{2}$	22	8	$32\frac{1}{2}$	$32\frac{1}{2}$	23	$187\frac{1}{2}$
56. Lancaster	$29\frac{1}{2}$	$42\frac{1}{2}$	$11\frac{1}{2}$	$5\frac{1}{2}$	9	53	15	15	181
57. Dallas	19	38	38	54	17	$5\frac{1}{2}$	$3\frac{1}{2}$	3	178
58. Flint	11	64	3	$12\frac{1}{2}$	$29\frac{1}{2}$	20	14	$17\frac{1}{2}$	$171\frac{1}{2}$
59. New Orleans	$12\frac{1}{2}$	4	46	30	24	14	16	14	$160\frac{1}{2}$
60. Ft. Worth	$22\frac{1}{2}$	$45\frac{1}{2}$	$24\frac{1}{2}$	35	18	$5\frac{1}{2}$	$3\frac{1}{2}$	2	$156\frac{1}{2}$
61. San Diego	8	25	36	43	6	18	9	$6\frac{1}{2}$	$151\frac{1}{2}$
62. Utica-Rome	$25\frac{1}{2}$	33	$15\frac{1}{2}$	3	2	$28\frac{1}{2}$	12	$10\frac{1}{2}$	130
63. Miami	2	29	5	39	15	10	9	8	117
64. Wichita	14	38	2	18	19	3	$3\frac{1}{2}$	$4\frac{1}{2}$	102
65. High Point-Greensboro	3	$14\frac{1}{2}$	9	4	14	$25\frac{1}{2}$	$6\frac{1}{2}$	$10\frac{1}{2}$	87

[a] From Air Eng., 9:9, 22–23, September 1967.

[b] 65 indicates most severe pollution.

[c] Paterson-Clifton-Passaic and Canton are tentatively listed below Milwaukee because it is believed that adequate sulfur dioxide data, not now available, will place them there.

[d] Lawrence-Haverhill position is tentative as a result of the estimation of gasoline factors.

TABLE 1-4

Summary of Air Sampling Data--1964-1965[a]

	Number of biweekly sampling stations	Concentrations in micrograms per cubic meter unless noted	
		Arithmetic Average[b]	Maximum
Urban Stations	291	105	1254
Suspended particulates	218	6.8	c
Fractions:			
Benzene-soluble organics	96	2.6	39.7
Nitrates	96	10.6	101.2
Sulfates	56	1.3	75.5
Ammonium	35	0.001	0.160
Antimony	133	0.02	c
Arsenic	100	0.0005	0.010
Beryllium	35	0.0005	0.064
Bismuth	35	0.002	0.420
Cadmium	103	0.015	0.330
Chromium	35	0.0005	0.060
Cobalt	103	0.09	10.00
Copper			

Iron	104	1.58	22.00
Lead	104	0.79	8.60
Manganese	103	0.10	9.98
Molybdenum	35	0.005	0.78
Nickel	103	0.034	0.460
Tin	85	0.02	0.50
Titanium	104	0.04	1.10
Vanadium	99	0.050	2.200
Zinc	99	0.67	58.00
Gross beta radioactivity	323	0.8 pCi/m^3	12.4 pCi/m^3
Sulfur dioxide	36	62	1100
Nitrogen dioxide	31	95	374
Nonurban Stations			
Suspended particulates	32	37	312
Fractions:			
Benzene-soluble organics	28	1.2	c
Arsenic	24	0.005	0.02

aFrom: _Air Quality Data--1964-1965_, U.S.P.H.S., D.H.E.W., Cincinnati, Ohio, 1966.

bArithmetic averages presented for comparison with previous data. The geometric mean for all urban stations was 90 μg/m^3, for the nonurban stations 28 μg/m^3.

cNo individual sample analyses performed.

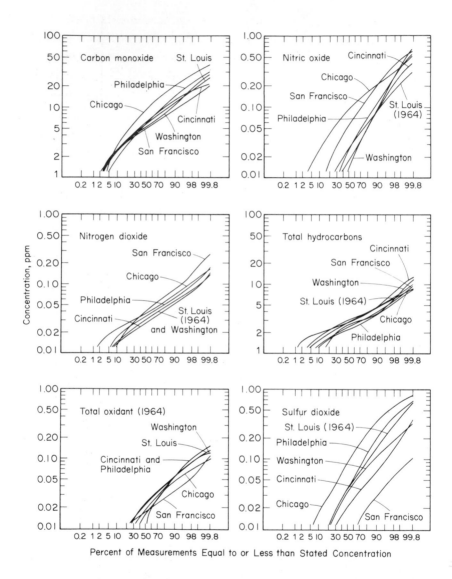

FIG. 1-2. Frequency plots from CAMP. From _Air Quality Data--1964-1965_, U.S.P.H.S., D.H.E.W., Cincinnati, Ohio, 1966.

that it may well dull the senses at levels considerably below those that have been believed toxic. The threshold limit value (TLV) for carbon monoxide is 50 ppm_v, lowered from 100 ppm_v several years ago, for occupational (40 hr/wk) exposure; CAMP and National Air Sampling Network (NASN) data show that the normal levels in our major cities are a significant fraction of the TLV, and measurements in tunnels and traffic jams have frequently exceeded this level. Smokers expose themselves to several hundred ppm_v while they are smoking. Sulfur dioxide should be placed on the most notorious list because it is high in obnoxious-ness and nuisance-causing potential and produces corrosion of materials and plant damage. Adverse health effects from sulfur dioxide at low concentrations have yet to be proved to the satis-faction of many workers in the field. Dirtiness remains a problem in urban areas despite the attempts to control air pollution; this may be the first urban air pollution problem to be solved, but attaining the standards being set in some cities is going to be exceedingly difficult. Pittsburgh shows quite dramatically the improvements that can be made if sufficient effort is expended. These problems appear destined to remain in the order named through at least the next several years.

C. Rural Problems

Agricultural dust has been mentioned among the general problems because it is so widespread. A large area that includes parts of Texas, Oklahoma, Colorado, and Nebraska has the worst problem with blowing dust. This semiarid grassland has been plowed up to plant wheat and other crops. Great clouds of dust from this area often travel hundreds of miles, producing visibility restrictions, choking, and general dirtiness all along their paths. Perhaps the air pollution problems caused by this nuisance will result in better control than have the soil conservation problems.

People in rural areas have tended to abuse the air resource with indiscriminate burning of their domestic refuse, field stubble, patches or piles of brush, sawmill wastes, cotton gin wastes, dead livestock, dry cattle manure, weeds (even green ones), and anything else that burns, with or without the aid of kerosene, distillate (tractor fuel), or gasoline. The population density now leaves few places where such burning does not affect neighbors or some town or city by the release of products that travel considerable distances with the winds before dissipation or removal.

Pesticides cause air pollution problems in many rural areas. Fields and even forests and lakes are sprayed by fogging or by dusting to obtain good distribution on plants and surfaces. Small particles remain airborne in the area and without dilution cause local problems. Wind carriage may cause problems at a distance; wind drift of herbicides has caused havoc among fields bordering the one being sprayed.

In addition to man-made (anthropogenic) air pollution in rural areas, nature adds quite a list of sometime pollutants. The principal natural air pollution, other than climatic, is usually the result of pollen. An acre with ragweed may release on the order of 10^{12} particles in a season; a few hundred per cubic meter cause problems to hay fever sufferers. Terpenes have received a lot of attention recently; these are hydrocarbon ($C_{10}H_{16}$) materials produced by pine trees and other evergreens. Marsh gas and radon also cause some rural problems, the latter for prospectors looking for uranium. The "smoke" of the Smoky Mountains was there before man.

D. Specific Problems

Some of the severe problems that have been alleviated if not cured must be cited in the interest of touching on the diverse history of air pollution. Other problems are everlasting it seems and only wait for the proper set of conditions to reappear.

A classic example of an air pollution problem which has gone full cycle is that of Ducktown, Tennessee. In the early part of this century, sulfur dioxide from copper smelting in Ducktown denuded the countryside of vegetation. Taller stacks were installed to solve the problem. They spread the problem over a much wider area of Tennessee and Georgia. The interstate nature of the problem brought federal action. Sulfur dioxide recovery was installed to remove the high concentrations.

Trail, British Columbia, was similar to Ducktown except that the problem was international, with many complaints from the downvalley region in Washington. Twice the problem was mediated by a neutral nation and damages paid to the farmers whose crops had been damaged. An operating program was set up which called for curtailment of operations during adverse weather conditions so that no more than a set tonnage of sulfur dioxide would be emitted from the stacks when poor dilution prevailed; even plant shutdown was required under the severest conditions. Later, sulfur dioxide recovery systems were installed.

Los Angeles, California, and its smog problems have received wide notoriety in air pollution circles. The controls already put on industry and automobiles have improved the hydrocarbon and carbon monoxide emissions but the oxides of nitrogen have increased and the smog goes on. The Los Angeles basin is one of the worst locations in the world for weather-aided air pollution problems. A strong subsidence inversion persists for long periods (weeks) and is accompanied by light onshore winds. The effect is comparable to a pot with a lid on it; all pollutant emissions are kept in the pot and stewing by photochemically-driven reactions.

Pittsburgh, Pennsylvania, and St. Louis, Missouri, have both tackled extremely severe smoke and particulate problems with enough zeal to alleviate the problems to a major degree.

These cities had midday palls of smoke which required street
lights to be lit. They still have air pollution problems but to a
much lesser extent than before.

The Dalles, Oregon, had a fluorides problem caused by an
aluminum extraction plant. The plant found it more economical
to change its process method rather than to remove the fluorides
from its off-gases.

New York, London, and several other cities have air pollution problems which have existed for a very long time and will
probably continue to exist for some time to come. At least one
of the mistakes continually made is the issuance of regulations
as to what must be burned rather than as to maximum allowable
emissions. Such an order by New York has resulted in an impor-
tation of several million tons of low-sulfur coal from West Ger-
many, which was probably not in the best interests of air pollution
control. Every 5-10 years these cities suffer periods of air
pollution which may well be called catastrophes. Some of these
are described in the next section.

E. Air Pollution Catastrophes

Most connections between air pollution and health effects
are tenuous and difficult to establish other than through long-
term, large-scale statistical correlations. There have been some
acute episodes which lasted a few days and were the undoubted
causes of many deaths. All these catastrophic occurrences had
one factor in common--they occurred during periods of stagnation
when pollutants were not carried away by the wind but accumulated in a relatively small volume of the atmosphere.

London has had two disastrous periods of air pollution in
the recent past. From December 5 to 9 in 1952, London suffered
a period of smog during which about 4000 deaths were attributed
to air pollution. This was the number of deaths above those that

would be expected during a like period without the smog. In the second week of December in 1962, London had another major episode of air pollution with the smog blamed for several hundred deaths.

The Meuse River Valley in Belgium had an air pollution catastrophe for the first 5 days of December in 1930. There were 63 deaths attributed to the pollution which accumulated under an inversion below the tops of the valley walls.

New York City has undergone at least two such experiences, the most recent in the fall of 1966. The New York metropolitan area contains in excess of 15 million people who are exposed to rather severe air pollution; according to Table 1-3 it has the worst city pollution in the United States.

All of the above incidents involved the presence of a multiplicity of pollutants in the air and no specific one could be cited as the cause for the deaths. The deaths were admittedly among the weak, the old, and the infirm, but they did occur.

Poza Rica, Mexico, had a catastrophe from a single pollutant, hydrogen sulfide. In November of 1950, a newly established industry released hydrogen sulfide and 22 people died. Several other incidents involving a single pollutant have required hospital treatment for a number of people exposed. These include fires in warehouses that stored parathion; releases of carbon monoxide, ammonia, chlorine, and so on, in transportation accidents and in process mishaps; and traffic jam exposures of policemen to carbon monoxide. Carbon monoxide deaths are numerous, but they occur in enclosed spaces.

The severest incident of fallout exposures to people occurred from one of the United States weapons tests in the Pacific. Rongelap natives were exposed to about 175 roentgens of radiation. This was sufficient to cause nausea and epilation. They also received beta burns of the feet and neck, but all of them recovered from this and from the whole-body dose. The exposure took place because a portion of the cloud sheared off and traveled in a different direction from the main cloud and the forecast direction.

F. How Much Air Pollution?

All the standard metropolitan statistical areas (SMSA) have some kind of air pollution problem; for example, noise is a serious problem in all of them; there are also problems to some degree with carbon monoxide and other auto exhaust products. The 212 SMSA's have a total area of 310,233 mi^2, or about 9% of all the land in the United States, and contain two-thirds of the population. Most of the remainder of the country has some kind of air pollution problem at least some of the time, involving either airborne materials or energy content.

TABLE 1-5

Inventory of Pollution Emissions[a]

Pollutant	Yearly emissions (Mtons)
Carbon monoxide	94
Natural dust	30
Oxides of sulfur	31
Hydrocarbons	29
Industrial dust and ash	22
Oxides of nitrogen	16
Other gases and vapors	2
Pollen	1.7

[a]National Emissions Standards Study, U. S. Senate Document 91-63, Washington, March 1970 and Other Sources.

It has been estimated that more than 200 million tons of air
pollutants are emitted in 1 year. Table 1-5 shows some of the
various pollutants and their estimated amounts. Note that carbon
dioxide, methane, and some other emissions, both natural and
man-made, are not on the list. Weight is a very poor measure of
pollution.

EXAMPLE 1-2: Are the particulate emissions in Table 1-5 unreasonably high?

In view of the many reported dustfall values of 60 $tons/mi^2$-
month, assume that 10 $tons/mi^2$-month is conservative as
an average for the SMSA's.

Dustfall on SMSA's = 310,000 mi^2 X 120 $tons/mi^2$-year
= 37 Mtons/year.

This leaves only 17 Mtons/year for the other 91% of the
United States, which is unreasonably low instead of high.
Also, the dustfall is measured several meters above the
ground, which lowers its amount considerably.

EXAMPLE 1-3: What is the approximate lifetime of airborne dusts?

Dust content of the air: 1957-1961 geometric means ($\mu g/m^3$)
were 21 for desert, 21 for mountains, 22 for forests, 34 for
farms, and 110 for cities. Assume that 35 $\mu g/m^3$ is the
average concentration to a height of 100 m and that negligible dusts are above that height.

Total weight = 35 X 10^{-9} kg/m^3 X 8.8 X 10^6 km^2 X 100 m
X 10^6 m^2/km^2 = 31 X 10^6 kg = 34,000 tons

If the assumed dustfall for the United States is 150-200
Mtons/year, the average lifetime is approximately 2 hr.

Long-term averages do not describe the extent of air pollution problems without some expression of the peak concentrations. There are fluctuations in concentration of pollutants,
some of them from degree of mixing (see Chapter 6, Section II, C)
and others by fluctuations in source emissions. Most pollutants
show systematic secular variations according to source: for
example, auto exhaust products levels peak at 8 a.m. and 5 p.m.
from Monday through Friday; space heating products are high in
winter; and soiling index (dirtiness of the air) follows industrial,
construction, automotive, and coal-burning activity (see Fig. 1-3).

28

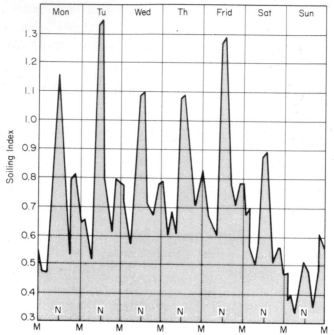

FIG. 1-3. Daily fluctuations of dirtiness. From "Troubled Air," U.S.P.H.S. Publ. No. 977, D.H.E.W., Washington, D. C., 1962.

VIII. SUMMARY

Air pollution reduces the usefulness of the air resource. Some substances considered pollutants can be found worldwide, namely, DDT and fallout. The increasing level of atmospheric carbon dioxide is considered by some observers to be a threat to coastal cities in the next few hundred to few thousand years because the heat absorbed by carbon dioxide could melt the polar ice caps. The most serious problems of our day are those in urban areas. The seriousness is based on the numbers of people who must tolerate the condition, as well as the severity of the concentration or level of the pollutant. The important urban problems, in order of importance, are noise, photochemical smog, pollen, carbon monoxide, sulfur dioxide, and dirtiness.

Rural air pollution problems include agricultural dust, pesticides, smoke and combustion products, and natural pollutants such as pollen, terpenes, marsh gas, and so on. Air pollution problems can be, and have been, rectified with the proper efforts. Serious sulfur dioxide problems at some copper smelters have been alleviated by sulfur dioxide recovery and tall stack dispersion. Smoke problems in Pittsburgh and St. Louis have been ameliorated by energetic programs of control. Air pollution catastrophes occur periodically because pollutants continue to be emitted after the atmosphere loses its capacity to dilute them because of periods of stagnation. Such episodes have resulted in numerous deaths but have not resulted in the elimination of the conditions for recurrence in periods of bad meteorological conditions. Perhaps the current increasing intolerance of people for air pollution will result in its abatement, provided the publicity-generated intolerance does not die.

PROBLEMS

1. Verify the weight of air on earth as shown on p. 2 (6×10^{15} tons).

2. Determine the composition of saturated air.

3. Convert 0.25 lb/1000 lb of flue gas at 50% excess (often abbreviated XS) air to milligrams per cubic meter.

4. Change the quantity in problem 3 to grains per cubic foot.

5. Assuming that there are 210×10^6 pollen particles per lb of pollen and making other suitable assumptions, calculate an estimate of the pollen concentration during season.

BIBLIOGRAPHY

Air Quality Data--1964-1965, U.S.P.H.S., D.H.E.W., Cincinnati, Ohio, 1966.

Community Air Pollution, U.S.P.H.S. Training Manual, Cincinnati, Ohio, 1961.

W. L. Faith, Air Pollution Control, Wiley, New York, 1959.

P. L. Magill, F. R. Holden, and C. Ackley, Eds, Air Pollution Handbook, McGraw-Hill, New York, 1956.

A. C. Stern, Ed., Air Pollution, Vol. I, 2nd ed., Academic Press, New York, 1968.

A Study of Pollution--Air, Staff Report, Committee on Public Works, United States Senate, Washington, D. C., 1963.

Waste Management and Control, National Academy of Sciences-- National Research Council, Publ. 1400, Washington, D. C., 1966.

Chapter 2

SOURCES OF AIR POLLUTION

I. INTRODUCTION

There are many and diverse sources of air pollution as
defined in Chapter 1. Man and his activities cause much of the
air pollution, but nature should not be overlooked as a source of
pollution. Table 2-1 shows an inventory of selected pollutants
and their sources. This table is a fair representation of what
has been emphasized in air pollution--the list is limited not
only to man-made pollution but also within that category to
little else than products of combustion. For example, it is
doubtful that the particulates in the transportation classification

TABLE 2-1

Sources of Some Man-Made Pollutant Emissions
in the United States (Percent of Pollutant)

Source	Carbon monoxide	Sulfur oxides	Hydro-carbons	Particu-lates	Nitrogen oxides
Transportation	92	4	65	14	42
Industry	4	32	26	44	21
Generation of electricity	--[a]	48	--	21	32
Space heating	3	12	3	14	5
Refuse burning	1	4	6	7	--

[a] -- = small amount.

31

include all of the following: dust from tires on road, rubber
particles from tires, exhaust particulates, photochemical smog
particles, and asbestos from the brake shoes of cars and trucks.
Obviously roadbed materials cast into the air are not included.
The omission of carbon dioxide was cited as deliberate in the
article that accompanied the table in the source, and such a
position is certainly defensible. Is it proper to include all the
carbon monoxide emitted by cars and trucks? According to the
definition of air pollution used here, carbon monoxide put into
the air in sparsely traveled wilderness areas hardly qualifies
as air pollution. However, the noise of transportation in urban
areas and along major highways and at airports should receive
the attention a severe problem merits. Odor is another pollution
ill that has been relegated to a position far below its degree of
seriousness, primarily because it is not measured in absolute
units meaningful to the engineer.

II. SOME RELEVANT COMBUSTION FACTS

An understanding of some of the factors involved in combustion and of the pollutants produced by combustion is worthwhile because combustion does cause many air pollution problems. Obviously, combustion uses up oxygen and produces
carbon dioxide in large quantities, but what is generally considered more important in air pollution is that particulates, oxides
of nitrogen, oxides of sulfur, and carbon monoxide result from
combustion. Also, hydrocarbons, water, acids, aldehydes, and
other substances which are intermediate and polymerized forms
of the substances entering the combustion process appear in the
off-gases in lesser amounts than those components listed above
but still cause many problems, probably the most significant
contribution currently being photochemical smog from hydrocarbons and polycyclic hydrocarbons. The high-velocity movements
of air associated with powerful burners make combustion a very
noisy process and it is notable for rockets, kilns, and so on.

The important combustion reactions are the following.

$$C + O_2 \longrightarrow CO_2 \qquad 14,100 \text{ Btu/lb (Higher Heating}$$
$$\text{Value)} \qquad (2\text{-}1)$$
$$H_2 + 1/2\, O_2 \longrightarrow H_2O \qquad 61,100 \text{ Btu/lb}$$

$$S + O_2 \longrightarrow SO_2 \qquad 4,000 \text{ Btu/lb}$$

$$H_2S + 3/2\, O_2 \longrightarrow SO_2 + 2\,H_2O \qquad 7,100 \text{ Btu/lb (hydro-gen sulfide)}$$

$$CH_4 + 2\, O_2 \longrightarrow CO_2 + 2\,H_2O \qquad 23,880 \text{ Btu/lb (meth-ane)}$$

$$C_2H_6 + 3\,1/2\, O_2 \longrightarrow 2\,CO_2 + 3\,H_2O \qquad 23,320 \text{ Btu/lb (eth-ane)}$$

$$C_3H_8 + 5\, O_2 \longrightarrow 3\,CO_2 + 4\,H_2O \qquad 21,660 \text{ Btu/lb (pro-pane)}$$

$$C_4H_{10} + 6\,1/2\, O_2 \longrightarrow 4\,CO_2 + 5\,H_2O \qquad 21,300 \text{ Btu/lb (bu-tane)}$$

$$NH_3 + 1\,3/4\, O_2 \longrightarrow NO_2 + 1\,1/2\,H_2O \qquad 72,620 \text{ Btu/lb (ammo-nia)}$$

The reactions shown are for complete combustion. For complete combustion to occur, the temperature, time, and oxygen supply must be sufficient, a situation that does not occur in practice. Combustion is a probabilistic process because the oxygen atom and the combining atom must be adjacent to each other for oxidation to proceed. To give a high probability of reaction, an excess of oxygen (air) is used and turbulence is encouraged in the burning chamber. The incomplete combustion leads to the discharge of the gaseous pollutants previously listed, plus some of the particles; other particles come from inert materials in the fuel which appear as ash. There are many reactions that take place during combustion and those above should be construed as being only the most important of the combustion reactions.

The farther the combustion proceeds toward completion, the less the amount of carbon monoxide and hydrocarbons emitted but the greater the amount of oxides of nitrogen formed from nitrogen in the oxygen source (air) (see Fig. 2-1). There-

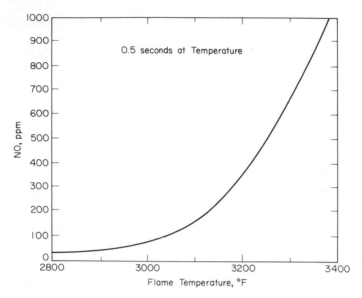

FIG. 2-1. Theoretical nitric oxide production versus
temperature. From J. D. Sensenbaugh and J. Jonakin, "Effect
of Combustion Conditions on Nitrogen Oxide Formation in
Boiler Furnaces," A.S.M.E. Paper 60-WA-334, 1960.

fore large, stationary combustion plants usually produce fewer
unburned or intermediate products but more oxides of nitrogen,
than smaller units. The automobile engine with its very high
turbulence, rather high temperature, and good oxygen-to-fuel
ratio results in relatively incomplete combustion because of the
short time available for combustion to occur.

III.   TRANSPORTATION

Nowhere does it seem more likely that man's ability to
pollute has outstripped his capabilities of pollution prevention
than in transportation. Advances in speed and power have
generally been accompanied by increases in noise and exhaust
pollution. Most of the exhaust pollution comes from incomplete
combustion despite the high-octane gasolines now being used.

A. Noise

Some of the most powerful rockets make noises plainly audible at 30 miles, but because of the remote firing sites only a relatively small number of people are exposed to rocket noise. More serious noise problems result from jet aircraft and from surface transportation.

Jet aircraft are reported to produce a noise of 140 dB, or 106 dBA, at 100 ft. The noise is caused by the relative movements of the turbine parts and by the exhaust shear. Although noise attenuation from a point source can be assumed to follow the inverse square law (twice the distance, one-fourth the sound), the decibel is so defined (Chapter 7, Section III, H) that the attenuation is 6 dB for lowering the intensity to one-fourth its former level. The large number of flights at major airports now require takeoffs and landings at 1-min intervals. This high frequency, coupled with the increasing population density around most airports, makes jet noise a very obnoxious problem. The sonic boom will become an important environmental pollutant with the advent of the supersonic transport (SST). The intensity of the boom, usually given in millibars, or pounds per square foot overpressure correlates directly with the airplane weight and inversely with the height to the 3/4 power. The tolerance level of people for sonic booms was tested by the U.S. Air Force and found to be low. A 1.7 psf overpressure boom was found to be as tolerable as 105 dB of subsonic noise. Some observers have predicted that the use of the SST's will be limited to transoceanic flights because of the expected complaints along any populated corridor.

Trucks, cars, and motorcycles emit noise primarily from their engines and tires and as a result of wind resistance, but sometimes horns and rattles also become significant noise sources. Engines and tires are the principal sources and are about equal in importance. The noisiest vehicles are those that lack mufflers or have ineffectual ones. Noise levels that have been reported omitted the noise from faulty mufflers, tire squeals from braking and accelerating, and horns. The levels measured at 50 ft and given for 10% noisier than, modal frequency, and 10% quieter than are: tractor-trailer trucks, 89, 86, and 81 dBA; dump trucks, 84, 82, and 78; motorcycles, about the same as for trucks; and cars (at 65 mph), 75, 72, and 69. If two noises are strictly additive, the resultant noise, which is twice the intensity of either source, is 3 dB higher than either noise alone. Modern traffic densities often result in rather intolerable noises along many freeways.

Train engines produce a noise level of about 96 or 97 dBA at 50 ft, while track noise is about 90 dBA. The noise from a string of four or more engines is attenuated as a line source for a distance up to 5 or 10 times the aggregate length. The line source produces one-half the noise at twice the distance or a 3-dB reduction for doubling the distance. Efforts are being made to reduce train noise and some of the techniques appear promising.

B. Exhaust Gases and Vapors

Carbon monoxide from transportation exhausts amounts to about 50% of the total weight of air pollutant emissions as classified by some workers. In addition, the hydrocarbons in exhaust gases constitute nearly 8% of the total. Exhaust gases, including oxides of nitrogen and sulfur, but not counting water vapor and carbon dioxide as pollutants, comprise about 60% of the total pollutants as shown in Table 2-1.

The approximate emissions from jet, auto, and diesel exhausts are shown in Table 2-2. Reference to this table reveals that the diesel engine produces much less carbon monoxide and fewer hydrocarbons but considerably more oxides of nitrogen

TABLE 2-2

Exhaust Emissions[a]

| Pollutant | Planes (lb/landing and take off) | | Autos[b] | Diesel |
	Jets	Piston	(lb/1000 gal fuel)	
Aldehydes (as formaldehyde)	2.2	1.1	4	10
Carbon monoxide	20.6	9.0	2300	60
Hydrocarbons (as carbon)	19.0	1.2	200	136
Nitrogen oxides (as nitrogen dioxide)	9.2	5.0	113	222
Particulates	7.4	2.5	12	110
Organic acids (as acetic acid)	--[c]	--	4	31
Sulfur oxides (as sulfur dioxide)	--	--	9	40

[a]From R. L. Duprey, Compilation of Air Pollutant Emission Factors, U.S.P.H.S. Publ. No. 999-AP-42, NCAPC, Durham, North Carolina, 1968.

[b]For autos at high altitude (Denver): hydrocarbons, up 30%; carbon dioxide, up 60%; nitrogen oxides, down 50%.

[c]Data not available.

and organic acids as a result of its higher-temperature, more efficient combustion. The higher particulate load in diesel exhaust comes from the unburned carbon, and the greater amount of sulfur dioxide is a result of the higher sulfur content of the fuel.

The exhaust hydrocarbons contain many olefins (particularly the low-molecular-weight mono- and diolefins), paraffins (straight-chain and branched), aromatics (mostly benzene and toluene), and acetylenes. More than 80 components have been identified, but those receiving the most attention are olefins and polycyclics--olefins for their role in photochemical smog formation and polycyclics because they have been shown to be carcinogenic (cancer-causing).

<u>EXAMPLE 2-1</u>: If the average city auto travels 25 mi per day and
emits 0.08 lb carbon monoxide per mile, what is the
volume of carbon monoxide emitted per car per day? What
volume of air is required per car per day to dilute the carbon monoxide to 30 ppm, assuming uniform mixing?

Carbon monoxide: $25 \text{ mi/day} \times 0.08 \dfrac{\text{lb CO}}{\text{mi}} \times \dfrac{1}{28} \dfrac{\text{lb mole}}{\text{lb}} \times$

$385 \dfrac{\text{ft}^3}{\text{lb mole}}$ at NTP $= 27.6 \text{ ft}^3/\text{car-day}$

Dilution: $27.6 \text{ ft}^3 \times 10^6 \dfrac{\text{ppm}}{\text{million parts}} \times \dfrac{1}{30 \text{ ppm}} = 10^6 \text{ ft}^3$

C. Particulates

The particulate air pollution from transportation comes
from exhaust emissions, wear of tires on the roadbed, and erosion
of the roadbed by tires and by winds generated in vehicular
motion.

Although the weight of the exhaust particulates is relatively
small (less than 2 million tons/year), considerable significance
is attached to these particles because of the presence of known
carcinogens and of lead. More than 24 polycyclic hydrocarbons
have been identified in exhaust particulates, the one of most
interest being benzo[a]pyrene because it is high in both concentration and carcinogenicity. The concentrations of benzo-
pyrenes, primarily benzo[a]pyrene, in California have been
reported at levels of $0.1-61 \text{ ng/m}^3$ in the cities and $0.01-$
1.9 ng/m^3 in rural areas. The amount of lead (as tetraethyl or
tetramethyl) added per gallon of gasoline has declined because
of the trend toward regular gasoline and the availability of
higher-octane fractions released by the changeover to jet aircraft; however, gasoline consumption has increased markedly
so that in the past 10 years the amount of lead used in gasoline
in the United States has increased by one-third to nearly 1
billion lb/year, or 250,000 tons. Maximum ambient urban concentrations of lead have ranged as high as $42 \text{ }\mu\text{gm/m}^3$, but the

average is less than 5 $\mu gm/m^3$. Some investigators have reported that 30 - 40%$_w$ of the lead particulates are greater than 5 μm in diameter, but others maintain that virtually all of the particles are less than 1 μm in size.

Dusts from brake shoes and clutch plates were given considerable attention because asbestos fibers have been indicted as carcinogenic and asbestos bodies were reported in widespread tests. There does not appear to be a serious problem from this source because recent tests have shown that very little asbestos fiber results (mostly dust), and the tests have been shown to give positive results for materials other than asbestos.

Large amounts of dusts result from the action of tires and wind on unpaved roadbeds. These dusts are given little attention--they have been present for a long time and are rather common in nature.

D. Odors

The strongest source of odors in transportation is diesel exhaust. The distinctive, pungent odors emitted by diesel engines have been attributed to aldehydes, but current research by Vogh (U. S. Bureau of Mines) indicates that the odor is not caused by lower aldehydes and perhaps not by aldehydes at all. The problem is compounded by the low level at which a city bus emits its exhaust.

The dioxides of sulfur and nitrogen contribute to exhaust odors, as do many hydrocarbons. Carbonyls and other oxygenated compounds probably cause some of the diesel odor. It is hoped that the specific odorous materials will soon be identified and quantified to facilitate control measures.

E. Other Emissions

Of the other emissions from transportation, the ones that have been of most concern are hydrocarbons evaporated from carburetors and gasoline tanks. Because of the role of hydro-carbons in the formation of smog, California has formulated regulations on the sources noted above, and the filling of car tanks at service stations and tanks at refineries, has caused concern. 1971 model cars have an activated carbon adsorption device for capturing the evaporations.

IV. INDUSTRY

Although the emissions inventory shows industry in a favor-able light, industry has caused many serious air pollution prob-lems. These problems have arisen because of the sizes and loca-tions of industries--if they were sufficiently small and well dispersed, the atmosphere could assimilate industrial wastes with little difficulty. Large industries can produce more economi-cally than small ones, and industries tend to locate in close proximity to each other and to population centers. Figure 2-2 shows typical industrial emissions for a large industrial city.

A. Commercial Combustion

The name "commercial combustion" as used here includes those plants in which burning is the prime or only process; that is, fossil fuel power plants and refuse incinerators.

1. Fossil Fuel Power Plants

About 80% of the electricity produced in the United States in 1968 came from the combustion of fossil fuels, about 10^{16} Btu of thermal energy. Approximately 70% of this heat came from coal, 25% from gas, and 5% from oil. Estimates show that the heat derived from each of these sources will almost double by 1980 and will at that time comprise an estimated 63% of the total power generation.

a. Natural Gas Combustion. Natural gas, except "sour gas" (gas containing hydrogen sulfide), burns with little produc-

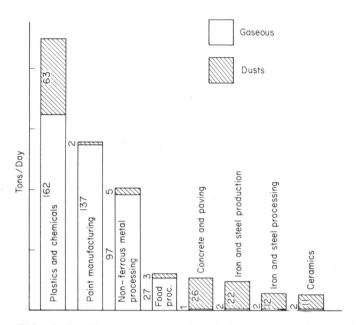

FIG. 2-2. Hypothetical industrial emissions in a major industrial city. Adapted from Troubled Air, U.S.P.H.S. Publ. No. 977, Washington, D. C., 1962.

tion of air pollutants. As a result, it is in great demand in areas fighting air pollution; however, the amounts of gas are carefully allotted to various areas by the Federal Commerce Commission (FCC). The amounts of various polluting materials in pounds per 10^6 ft^3 of gas (about 10^3 Btu/ft^3) have been listed as: oxides of nitrogen (as nitrogen dioxide), 390; oxides of sulfur (as sulfur dioxide), 0.4; aldehydes (as formaldehyde), 1; other organics, 3; and particulates, 15. Carbon monoxide and hydrocarbons are negligible for proper combustion. The emissions for smaller industrial boilers and the like are somewhat higher except for the oxides of nitrogen, which are lower, and the oxides of sulfur which are unchanged.

    b. Fuel Oil Combustion. Fuel oils used for power plants and industrial purposes are residual oils rather than distillates because residuals are cheaper. Pollutant emissions from fuel oil combustion are reported in pounds per 1000 gal of oil (about 8.2 lb/gal) as: sulfur dioxide, 160 X percent sulfur [in No. 6 fuel oil sulfur content is usually $1.6\%_w$ (range 0.34-4)]; sulfur trioxide, 2.75 X percent sulfur; oxides of nitrogen (as nitrogen dioxide), 104 (range 68-156); particulates, 8 (range 6.7-14.8); hydrocarbons, 3.2; aldehydes (as formaldehyde), 0.6; and carbon monoxide, 0.04. As is the case for gas combustion, smaller industrial boilers and so on burn at lower temperatures and therefore show increases in all the emissions except the oxides of nitrogen (lower) and the oxides of sulfur (same).

The particulates are made up by weight of 10-30% ash, 17-25% sulfates, and 25-50% cenospheres (burned-out skeletons of fuel particles). The size of the particles is quite variable with reports of 95% < 0.5 μm; 100% < 4 μm and 47% < 3 μm; 53% > 4 μm; and 95%, 10-1000 μm. It seems probable that the particle size reported depends upon not only the combustion process parameters (temperature, excess air, time of residence, and so on) but also the method of collection, with its attendant degree of breakage of the fragile, flocculant particles. Perhaps, saying that $50\%_w$ < 5 μm provides as good a description as is possible at the present time.

c. Coal Combustion. Nearly 300 million tons of bituminous coal were burned by utilities in 1968 for the generation of electricity, and an additional 200 million tons were burned by other industries. This coal had $10-15\%_w$ ash and contained about $2.5\%_w$ sulfur (range 0.5-4.5). Anthracite (hard) coal is low in sulfur ($0.4-0.8\%_w$) but is available only in limited places and amounts. Practically all of the sulfur comes out as sulfur dioxides, which means that the emissions are about 3.8 lb of sulfur dioxide per 10^6 Btu (13,000 Btu/lb of coal).

Coal contains some chlorine, about $0.1\%_w$ (range 0.01-0.46). In addition, $0.1-0.5\%_w$ calcium chloride is sometimes added to the coal as an antifreeze or dust control agent. With a reported $60\%_w$ emission as hydrogen chloride, this means the emission of 0.08-0.3 lb of hydrogen chloride per 10^6 Btu.

The other gaseous and the particulate emissions from coal burning vary with the combustion process used and are categorized in Table 2-3. The portion of the ash that entrains in the flue gas, that is, the fly ash, ranges from $10\%_w$ for the cyclone to $85\%_w$ or more for the dry bottom (the temperature of the ash in the hopper is below its melting point, while wet bottom indicates that it is above and the ash is in the form of slag) pulverized method of combustion.

EXAMPLE 2-2: Determine the air requirements (20% XS) and the flue gas composition from burning a bituminous coal with a weight percent analysis of: carbon, 63.50; hydrogen, 4.07; sulfur, 1.53; oxygen, 7.46; nitrogen, 1.28; water, 15.00; ash, 7.16 (HHV = 11,350 Btu/lb). Use 1 lb of coal for calculations. Make calculations in pound moles.

Component	Pounds	Moles	Moles oxygen	Moles product	
Carbon	0.6350	0.05292	0.05292	0.05292	carbon dioxide
Hydrogen	0.0407	0.02035	0.01018	0.01018	water
Sulfur	0.0153	0.00048	0.00048	0.00048	sulfur dioxide
Oxygen	0.0746	0.00233	-0.00233	--	
Nitrogen	0.0128	0.00046	"inert"	0.00046	nitrogen
Water	0.1500	0.00833	--	0.00833	water
		Totals	0.06358	0.07237	
			-0.00233		

Air required: $[1.20(0.06358) - 0.00233]\ 100/21 = 0.3523$ mole

Weight: $0.3523(29.1) = 10.25$ lb

Volume: $0.3523(385) = 136\ \text{ft}^3$

TABLE 2-3

Emissions from Coal Combustion[a]

| Combustion unit | Particulate | | | Gaseous (lb/10^6 Btu) | | | | |
	Amount (lb/10^6Btu)	Size M_g	σ_g	Nitrogen oxides	Aldehydes (formaldehyde)	Carbon monoxide	Hydro-carbons	Sulfur dioxide
Pulverized[b]	8	20	3.6	0.8	2×10^{-4}	0.015	0.008	1.6 X $\%_w$ sulfur
Cyclone	1	4	8	2.4	1.7×10^{-4}	Trace	Trace	1.6 X $\%_w$ sulfur
Spreader stoker	6.5	70	4.7	0.7	0.6×10^{-4}	0.03	0.009	1.6 X $\%_w$ sulfur
Hand-fired	1	--	--	0.3	2×10^{-4}	2	0.4	1.6 X $\%_w$ sulfur

[a]From W. S. Smith, Atmospheric Emissions from Coal Combustion, U.S.P.H.S. Publ. No. 999-AP-24, NCAPC, Cincinnati, Ohio, 1966.

[b]General, dry bottom, and wet bottom without fly-ash reinjection. Fly-ash reinjection increases the particulate load to about 12 lb/10^6 Btu.

Component	Moles	Percent by volume	Parts per million by volume
Carbon dioxide	0.05304^a	14.63	146,300
Water	0.01851	5.11	51,100
Sulfur dioxide	0.00048	0.13	1,300
Oxygen	0.01272	3.50	35,000
Nitrogen	0.278^b	76.63	766,300

[a] $0.05292 + 0.3523(0.00033) = 0.05304$

[b] $0.00046 + 0.3523(79/100) = 0.278$ (includes argon, nitrogen oxides, and trace gases)

A good rule of thumb on stack gas is that 50% excess air produces about 12% carbon dioxide, 7% oxygen, and 81% nitrogen.

2. Refuse Incinerators

Refuse is incinerated to reduce the volume of solid wastes and to break down putrescible organics into biologically stable inorganics. In the United States the custom is to waste the heat derived from incineration ($\sim$4000 Btu/lb) by burning the refuse with 250% excess air. The large amount of air keeps the temperature down to prevent damage to the refractory lining and to help ameliorate slagging problems. In Europe, where the heat is likely to be used, combustion is carried out with about 50% excess air, and water-carrying boiler tubes keep the refractory from overheating.

The pollutants emitted by efficient incinerators depend upon how much of the material burned degrades into polluting substances. Oxides of nitrogen are not much of a problem because of the low combustion temperature. Sulfur dioxide is usually not too serious because sulfur constitutes only about $0.1\%_w$ of refuse and Kaiser says that much of that goes into the ash as sulfates. There is increasing concern about the products from incinerating plastic materials, particularly chloride and fluoride-containing plastics. Incineration temperatures are sufficient to break down

the organic acids present in open burning, and carbon monoxide
is at a low level in flue gases.

The particulate emissions that are about 20 lb/ton of refuse
have been a vexing problem but the problem is being solved by
the advent of routine air-cleaning equipment on incinerator
effluents. Poorly operating incinerators and frequent start-ups
have resulted in incomplete combustion of organic odors and produced many complaints. The size of particles has been reported
as $40\%_W > 44$ μm and $15\%_W < 5$ μm.

B. Rock and Solid Mineral Products

Some air pollution causes are common to all rock and
solid mineral industries, while others are limited to a given
industry. One or more of the processes of quarrying and mining,
crushing and milling, classification and separation, heating or
roasting, grinding and mixing, and handling and transport apply
to the production of cement, lime, wallboard (sheet rock), glass,
ceramics, fiber glass, rock wool, hot-mix asphalt concrete,
perlite/vermiculite, building stone, riprap, aggregates, ores and
ore concentrates, and coal. Figure 2-3 is a flow sheet for these
operations. In addition to the process releases, piles of raw
materials or process wastes are often eroded and dusts entrained
in the wind to cause problems.

Quarrying and mining in rock or other hard materials involve
drilling and blasting. Drilling is usually done with a pneumatic
hammer using an air-cleaned or a water-cleaned bit. An air-
cleaned bit gives rise to considerable airborne dust. Blasting
often causes a large dust cloud, especially when the surface
is dry and dusty. The gases, vapors, and smoke from exploding
TNT cause headaches in those who inhale them before they are
diluted by travel. After the material is loosened, it is carried to
a crusher for size reduction. The transport may be carried out

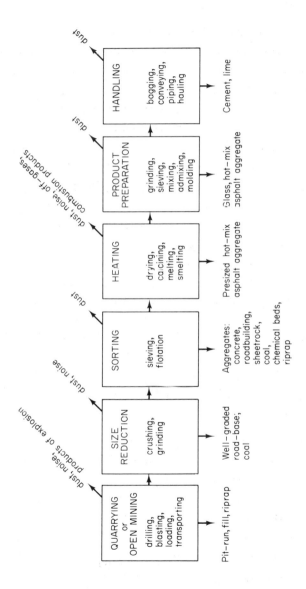

FIG. 2-3. Process flow sheet for rock and solid mineral industries.

by truck, train, or conveyer belt--any one of which may result
in the release of dust into the atmosphere.

The crushed material may be pulverized into small granules
and separated (often by screening or gravity, with or without
water) into various components, or sorted by size. The acceptable material then enters the process, which in the rock and
solid mineral industries is usually heating. This heating may
be for the purpose of drying, roasting, calcining, melting, or
for process reasons. Heating can take place in any of several
types of ovens but is most often done in rotary kilns. The kilns
are usually long drums placed nearly horizontal with just enough
slope to move the material through the drum as it is picked up by
fins inside the drum, carried well up toward the top of travel,
and dumped through the hot gases passing through the kiln.
The noise of combustion from the kiln can frequently be heard at
distances of 2 or 3 mi and is causing concern among plant oper-
ators of units located near population centers.

After heating, the material may be molded, ground, or mixed
into the final product. Grinding gives rise to more dust and another
size separation is usually carried out, causing even more dust
emission. The finished product is then packaged or shipped in
bulk.

1. Cement

About 400 million bbl (1 bbl = 376 lb) of Portland cement
were produced in the United States in 1967 at nearly 200 plants.
The plants range in production size from 0.675 to 13.8 million
bbl annually, with more than half of them in the 2- to 4-million-
bbl category.

About 600 lb of raw material containing lime (limestone,
oyster shell, chalk, or coral), silica (sandstone, shale, or
sand), alumina (clay or shale), and iron (iron oxide from ore)

are used to produce 1 bbl of cement; 35%$_w$ of the raw material
is driven off as carbon dioxide and water vapor. The material
must be low in magnesium. There are two processes used to
produce cement--the wet process ($\sim$60%$_w$ of total production)
and the dry process. In the wet process the ground raw materials
are mixed into a slurry with 30-45%$_w$ water. In the dry process
the kiln feed material is ground and dried to less than 1%$_w$ moisture
before introduction to the kiln. A rotary kiln is generally used
and may be 6-25 ft in diameter and 60-800 ft in length. The hot
gases of combustion move counter to the flow of process material.
As the material moves through the kiln, it is dried, decarbonated,
and calcined, then fused into clinker as it reaches a temperature
of about 2700°F. The clinker is ground to an average size of
about 10 μm, and ground gypsum is added to control the rate of
setting. Other additives are sometimes used.

The pollutant emissions from the process are primarily
particulates from the kiln and from the raw material production
preparatory to entering the kiln. The combustion products from
1 bbl of cement are those from burning 92.3 lb of coal or 8.27
gal of fuel oil or 1140 ft^3 of natural gas (1.2 x 10^6 Btu/bbl).
At least one cement plant has found it necessary to pave its roads
from the quarry and within the plant area to meet stringent new
property-line emission standards on dusts. When a drier is
used before the kiln in the dry process, the off-gases have dust
loadings of 5-10 gpscf, with higher values for the use of hot
kiln gases. The dry process gives off-gases from the kiln at
an average rate of 45,000 scfm per 1000 bbl/day capacity (range
12,000-80,000 cfm). These gases have a temperature of about
500°F (range 150-845°F) and contain an average 6.4 gr/ft^3
(range 1.1-12.4) of dust or a dust load reported at 46 lb of dust
per barrel of finished cement. Similar values for the wet process
are 45,000 cfm per 1000 bbl/day (22,000-68,000 cfm) at a
temperature of about 350°F (285-550°F) with a dust load of 5.7 gr/

ft^3 (0.995-13.5) or about 38 lb of dust per barrel of cement.
The kiln dust amounts to about $10\%_w$ of product weight and has a
size distribution as shown in Table 2-4. This dust is not cement.

2. Lime

Limestone is calcined in a kiln to produce calcium oxide
(CaO, quick lime). The calcium oxide may be marketed as such
or it may be slaked to calcium hydroxide, or slack lime. The
lime is ground to specified size on a nearly custom basis for
the customer, but the size is typically about $90\%_w$ passing a
200-mesh sieve. The fine material is usually removed from the
grinder by air; large particles are eliminated by a centrifugal
separator; the product is removed in a cyclone or series of
cyclones; and the cyclone exhaust is sent back to the grinder.
This apparently closed process picks up air which must be bled
out of the system, carrying with it considerable amounts (often
as much as $2-3\%_w$ of the process weight) of the fine materials.
(See Table 2-5 for the particle size of dusts.) Efforts are being
made to use sieve classification to avoid the air pollution caused
by materials in bleed air.

TABLE 2-4

Size Distribution for Particulate Emissions by Wet Cement Kilns[a]

Size (μm)	Percent less than
60	97-100
50	95-100
40	85-95
30	70-90
20	50-70
10	30-55
5	30-40
2.5	10-35

[a]From G. Nonhebel, Gas Purification Processes, George Newnes,
London, 1964.

TABLE 2-5

Size Distribution for Lime Particle Emissions[a]

Size (μm)	Percent greater than
44	28
20	66
10	90
5	98

[a]From R. L. Duprey, Compilation of Air Pollutant Emission Factors, U.S.P.H.S. Publ. No. 999-AP-42, NCAPC, Durham, North Carolina, 1968.

The lime may be sacked or shipped in the bulk. Practically no pollution other than the particulates comes from lime manufacturing, the only other products being carbon dioxide from the calcining and the products of combustion from the kiln. The particulates result from quarrying, crushing, burning, grinding (slaking), and handling. Wind entrainment of dusts from several acres of waste piles is causing a major problem at a lime plant of our acquaintance.

3. Glass Manufacture

Soda lime glass manufacture accounts for about $90\%_w$ of the total production of glass. In this process sand, limestone, soda ash, broken glass, and minor constituents are melted in a furnace at about $2700^\circ F$ to make glass.

The emissions from glass production, which are mineral dusts and fumes, amount to 2 or 3 lb/ton of glass in the melting and molding processes. Fluorides which are often in the feed materials go into the glass ($\sim 80\%_w$) or are emitted from the process ($\sim 20\%_w$). The particle size distribution has been reported at $1\%_w$ in the 20- to 44-μm-diameter range, $19\%_w$ 10-20 μm, $55\%_w$ 5-10 μm, and $25\%_w$ <5 μm.

4. Rock Wool and Fiber Glass

Rock wool and fiber glass are produced by melting the mineral ingredients ($\sim 3000^{\circ}$F) and blowing or drawing the molten material into small fibers a few microns in diameter. The fibers are usually made into mats for heat or sound insulation or filters or for fiber glass; or they may be spun into threads and made into cloth.

Particulate emissions from the furnace are principally coarse fumes, about 5-20 lb/ton, with a size distribution of $60\%_w$ > 44 μm, $27\%_w$ 20-44 μm, $10\%_w$ 10-20 μm, $2.5\%_w$ 5-10 μm, and $0.5\%_w$ <5μm. Emissions from the fiber processing may amount to $1\%_w$ of the process charge and consist of fiber particles $\sim$5 μm in diameter and often up to 0.5 in. in length. Considerable health significance is being attached to these airborne fibers and has resulted in major control efforts. Some batches of fiber glass do not meet specifications (are "sour") and are wasted. Wind erosion of discarded batches at a fiber glass plant in Central Texas caused complaints from physicians and their patients and resulted in lawsuits.

5. Hot-Mix Asphalt Plants

Hot-mix asphalt plants usually blend three to five sizes of aggregates to produce a well-graded aggregate which is then dried in a rotary kiln and coated with hot asphalt in a pug mill. Such plants have gained wide notoriety for their air-polluting capabilities. Many of the plants are portable and stay on a job site for a few weeks and then move to another. Before the strong efforts to control air pollution were initiated, hot-mix plants often discharged as much as a few percent of their process weight through a 25- to 50-ft stack. A simple cyclone precleaner now on most plants lowers the emissions to about $0.25\%_w$. The particulate size distribution emitted is approximately $3\%_w$

>44 μm, 20%$_w$ 20-44 μm, 17%$_w$ 10-20 μm, 25%$_w$ 5-10 μm, and 35%$_w$ <5 μm.

In addition to the dust problem, a problem now usually solved with scrubbers, hot-mix plants located near population areas receive many complaints about noise. The burners used in the drying kiln produce a roar which can frequently be heard for considerable distances.

6. Other Mineral Industries

There are many other rock and mineral industries that are not detailed here. Emissions from some of these other industries are delineated in pertinent references listed in the bibliography. Those for which emission descriptions cannot be found in the literature often have emissions problems predictable from a knowledge of the operation and comparison with similar delineated processes.

Sheet rock or gypsum wallboard production involves crushing gypsum (calcium sulfate rock) to a small size, adding moisture to bind, and sandwiching a layer of the material between two papers. Asbestos may be produced in a similar manner, except that no paper layers are used and the severest emission problems involve fibers rather than common dusts.

Vermiculite (perlite) is sort of a mineral popcorn which is mined in a compacted form and then expanded with heat for use as packing material or a lightweight concrete aggregate. Concrete batching plants frequently have dust problems from aggregate and cement handling and spills and from aggregate stockpiles.

Other dusty mineral industries are producers of poultry grit, asphalt shingle aggregates, glazing frit, and mica products. Chemical industries often use mined minerals as process feed and have the problems that have been described here.

C. Metals Industries

The production of metals starts with the ore. Some of the
same processes described in Section IV, B apply to the production
of ore, namely, mining, milling, and concentrating. Common
metals are separated from their ores by smelting and are then
purified by further treatment. Often the purification is done by
burning out the impurities. Other metals are separated electro-
lytically from solutions or molten ores.

The large amounts of air usually introduced into metal
production and purification entrain considerable amounts of
gaseous and particulate pollutants, making metals production
one of the dirtiest industries. Air pollution controls require a
large amount of effort because of the very small particulates given
off as fumes.

1. Iron and Steel

Iron- and steel-making industries have historically caused
many air pollution problems in Pittsburgh, Birmingham, and
other areas. The 1967 production of steel in the United States
was 127,213,000 tons. The character of the production process
is changing with the trend toward basic oxygen and away from
open hearth and Bessemer converters.

a. Blast Furnaces. Iron ore, with up to $60\%_w$ of iron, is
placed in a blast furnace in layers with coke, limestone, cinders,
and scale. For 1 ton of iron, the approximate amounts of materials
charged are 2 tons of ore, 1 ton of coke, 1/2 ton of limestone,
and 1/8 ton of scale and cinder. Combustion is supported by
about 4 1/2 tons of hot air. For blast furnaces that produce
1000 tons of iron per day, the volume of off-gases has been
reported at 87,000 scfm, containing 7-10 gr/scf (150 lb/ton of
steel) of particulates that range in size from 0.1 to several hundred μm ($68\%_w > 44$ μm). The off-gas contains about $25\%_w$

carbon monoxide which must be burned before the gas is released. The particulate removal is accomplished prior to flaring or burning the gas in a recovery furnace.

b. Open Hearth. Scrap iron and steel and pig iron are melted in a shallow furnace and the impurities removed in a slag and by burning in the open hearth process. Gases from the process are principally combustion gases and cause little pollution concern unless the fuel contains considerable sulfur, or unless fluorspar is used to condition the slag; however, the gases ($\sim 20,000$ scfm for a 100-ton furnace) contain 0.2-1 gr/scf of particulate ($5\%_w > 44$ μm and $45\%_w < 5$ μm) depending upon the phase of the cycle, the highest amounts occurring at charging and near the end of the heat. Oxygen lancing increases the emission rate about 50% but shortens the process and thereby reduces the overall emissions about 25%. The gas from the furnace has a temperature of 1200-1500°F and carries nearly all of the fuel sulfur as sulfur dioxide and about one-half the fluorspar as fluorides. The open hearth process is giving way to the much faster basic oxygen process.

c. Basic Oxygen. The basic oxygen process, also called the Linz-Donawitz or L-D process, uses a stream of oxygen directed at the surface of the hot metal to agitate the liquid and to promote combustion of the impurities, particularly carbon. The off-gases during the blow are 20 to 25 times the oxygen flow rate, which is often about 10,000 cfm for a 100-ton furnace. A blow may last 20 min and give off 200 lb of particulate (25-$50\%_w < 1$ μm), mostly iron oxide.

d. Electric Arc Furnace. Electrical resistance of the furnace charge is used to heat the metal. The 30,000 scfm (average) gas flow from a 50-ton furnace contains about 0.4 gr/scf (0.1-6) of particulates with $70\%_w < 5$ μm. The particulate often con-

tains large amounts of zinc oxide fumes along with the iron
oxide because of the zinc content of the scrap.

   c. Allied Operations. In addition to the basic processes
of iron and steel production, industry has other processes that
produce considerable amounts of air pollutants. These other
processes include coke production, sintering, and scarfing.

   Coke is produced by distilling the volatiles from crushed
coal; the coke, the remaining fraction, is nearly pure carbon.
The volatiles contain tar, benzene, naphtha, methane, hydrogen,
and hydrogen sulfide. Beehive ovens wasted the volatiles and
produced many air pollution problems in doing so. Modern by-
product ovens can produce coke with little air pollution, the most
difficult problem being the particles released in the quenching
process.

   Sintering is used to change the fines in ore to a cake to
prevent their entrainment in the blast furnace off-gases. Ore
is mixed with crushed coal which burns and gives the cake.
The flue gas (140,000 scfm for 1000 ton/day) contains 0.5-3
gr/scf of particulates ($90\%_w$ > 10 μm) and the cake discharge
has an additional 17,500 scfm gas stream with 6 gr/scf of large
particulates.

   A scarfing machine uses oxygen to burn clean the hot
surface of the billets and slabs before rolling. A 45-in., four-
sided machine is reported to give 85,000 scfm of off-gas with
0.2-0.8 gr/scf of particulates.

   2. Copper

   Copper production in the United States is about 2 million
ton/year and comes from sulfide ores. These ores are low grade
when mined and require concentration by gravity and/or flotation.
Ore is calcined by roasting and sulfur comes off as sulfur dioxide
(about 19 lb/ton of ore). The calcine is fed to a reverberatory
or a blast furnace for smelting and removal of most impurities as

slag. The copper moves to a converter for blowing with air to drive off remaining sulfur as sulfur dioxide and to oxidize iron for slagging. The copper is then refined by fire or electrolytic means. These processes give off carbon monoxide, oxides of sulfur, oxides of nitrogen, and small particulates (dusts and fumes). The off-gases have 1-6%$_v$ sulfur dioxide. The fumes released are copper, arsenic, antimony, lead, and zinc.

3. Lead

Lead ore, containing both lead and zinc, is put through concentration by differential flotation. The ore concentrate is roasted, sintered, and smelted in a blast furnace with coke and limestone. Sulfur dioxide emissions average about 550 lb/ton of ore. Lead fumes and dusts are entrained in the off-gases. One report shows 300 tons/day of particulate collected in baghouses treating 800,000 scfm of air (about 3.6 gr/scf).

4. Aluminum

Bauxite ore is processed, then calcined, to give alumina (Al_2O_3). Up to 1%$_w$ of the fine powder is emitted. Most aluminum production is carried out by electrolysis of the alumina in molten cryolite (3 $NaF-AlF_3$). About 89 lb of gaseous and particulate (equal amounts) fluorides are released during the electrolytic reduction of 1 ton of aluminum. The fluoride particles are quite small (0.05-0.75 μm) but are accompanied by considerable quantities of dusts which are coarser.

5. Other Metals Industries

The production of other primary metals is not deemed significant enough for coverage here, but the bibliography lists some publications that describe these other processes.

Foundries have caused many in-plant and ambient air pollution problems. Ordinary foundry operations result in a few pounds or even several pounds of fume per ton of molten metal. The fumes are small in size, mostly <1 μm. The loss of foundry sand (0.3 lb/ton of sand) causes industrial hygiene problems in the shake-out areas. These dusts are often collected and exhausted without control.

D. Chemical Industries

There is an almost endless variety of chemical industries. Only a few of them are selected for discussion here. Many of the air pollution problems from chemical industries are a result of the fact that plants are set up to process concentrated streams and waste the dilute streams, the gaseous to the atmosphere. This is particularly true of those materials that are high in obnoxious quality and not too high in price. Additionally, chemical reactions are often carried out at high pressures and temperatures, increasing the probability that vapors, gases, and so on will be released.

Complete sealing of a chemical process is virtually impossible. Leaks develop around packings in valves, pumps, and connections. Process changes resulting in dangerously high pressures or temperatures are frequently vented to the atmosphere on the premise that such is likely to occur not at all or infrequently. Considerable materials are usually wasted to the atmosphere during process start-ups and shutdowns.

Chemical plants are often disturbingly noisy. The flow of materials through pipes or by conveyor systems can be quite noisy when the transport is designed for efficiency and not for noise suppression. Combustion for a plant's power is normally very noisy. Steps are being taken to eliminate the high noise levels in new plant designs.

1. Acid

The manufacture and concentration of acid result in the release of process material and the finished product, acid mists.

a. Sulfuric Acid. Approximately 25 million tons of sulfuric acid were produced in the United States in 1967, about 90%$_w$ by the contact process and 10%$_w$ by the chamber process. The contact process uses vanadium pentoxide as a catalyst to convert sulfur dioxide to sulfur trioxide, which is then absorbed in 98-99%$_w$ sulfuric acid solution and makes the acid stronger. The chamber process uses oxides of nitrogen for its catalyst and produces an acid strength of only 60° Bé (77.7%$_w$) after a concentrating step from 50 to 54° Bé. The sulfur dioxide feed is in a gas stream of 8-11%$_v$ sulfur dioxide from an elemental burner, or 7-14%$_v$ from a pyrite roaster, or as low as 3%$_v$ from some processes. About 70%$_w$ of United States production is from burning elemental sulfur. The emissions from the contact process have less than 2%$_v$ (1.5-4) sulfur dioxide unconverted (~0.2%$_v$ sulfur dioxide in exit gas), unabsorbed sulfur trioxide at an average 13 mg/scf (0.5-48), and acid mist at 0.3-7.5 lb/ton sulfuric acid without a mist eliminator, or 0.02-0.2 lb/ton sulfuric acid with an eliminator. Sulfur trioxide emissions in any significant amount cause a visible plume. The acid mist that escapes removal is composed of small particles, 98%$_w$ < 3 μm. Other emissions are in the parts per million (by volume) range and include nitrogen oxides from combustion.

The chamber process has unconverted sulfur dioxide at 0.1-0.2%$_v$ when using elemental sulfur and 0.2-0.4%$_v$ for other sulfur sources, oxides of nitrogen 0.1-0.2%$_v$ (as nitrogen dioxide), and acid mist at concentrations of 0-30 mg/scf.

b. Nitric Acid. Nitric acid production of about 6 million tons in 1967 was carried out by the oxidation of ammonia to nitric oxide with a platinum-rhodium catalyst. The nitric oxide is cooled and combined with oxygen to give nitrogen dioxide (in equilibrium with N_2O_4). The nitrogen dioxide is cooled and absorbed in water to form an acid solution that is 50-60%$_w$ nitric

acid plus 1 mole of nitric oxide for 2 moles of acid. The nitric
oxide is returned to the process.

The off-gases of about 85 scfm/ton of acid per day contain
oxides of nitrogen (except nitrous oxide) at about $0.35\%_V$ $(0.1$-
$0.69)$ and small amounts of acid mists. Nitrogen dioxide (about
0.13-$0.19\%_V$) causes the orange color in the plume. The tail
gas may be treated with catalytic combustion to remove about
$80\%_V$ of the nitrogen oxides, or with alkaline scrubbers about
$90\%_V$.

c. Phosphoric Acid. Phosphoric acid is made by two processes, the wet and the thermal. The wet process is used mostly
for fertilizer manufacturing and consists of reacting sulfuric acid
with phosphate rock to give phosphoric acid and gypsum. The
gypsum and other insolubles are removed by vacuum filtration.
The acid is concentrated by evaporation to 50-$55\%_W$ phosphorus
pentoxide, or superphosphate is concentrated to 70-$85\%_W$ phosphorus pentoxide. The thermal process is used to manufacture
a purer grade of acid for food and chemical industries. This
process produces elemental phosphorus which is mixed with air
to give phosphorus pentoxide and scrubbed in water to form the
acid.

The principal emissions from the wet process include the
acid mists and gaseous fluorides (mostly silicon tetrafluoride
with some hydrogen fluoride), the latter at about 20-60 lb/ton of
phosphorus pentoxide produced. The thermal process gives acid
mists which after control range from 0.2 to 10.8 lb/ton of phosphorus burned. The particle size of the mists is reported to be
0.4-216 μm, with a mass median diameter of 1.6 μm.

Acid concentrators are used on nitric acid, phosphoric
acid, and other acids manufactured in strengths less than those
desired. The nitrogen oxide emissions from a nitric acid con-

centrator for making 98%$_w$ acid from 60%$_w$ have been reported as 5-7 lb/1000 lb of nitric acid.

2. Fertilizer

Fertilizer formerly was primarily animal manure or guano but is now mostly chemical or synthetic. Manure fertilizer causes odor problems, especially when it rains, and dust problems from transport and application. Chemical fertilizers may be granular materials high in nitrogen, phosphorus, and potassium, or the purer forms of anhydrous ammonia, superphosphates, and potash.

Air pollution associated with fertilizer production includes fluorides, acid mists, and dusts associated with phosphate production (see Section IV, D, 1) and carbon monoxide and ammonia emitted from ammonia manufacture. Ammonia is produced by re-forming natural gas with steam and catalytically reacting the products with nitrogen at high pressure. Each ton of ammonia produced by a 450-ton/day plant reportedly releases 7 lb of ammonia and 200 lb of carbon monoxide in a total stream of 1200 scfm.

Many fertilizer plants built without air pollution controls are experiencing difficulties in meeting stringent new air pollution standards. The major problems involve sulfur dioxide, ammonia, acid mists, fluorides, and dusts.

3. Carbon Black

In 1965 more than 2 1/4 billion (10^9) lb of carbon black (also called lamp black and black) were produced in the United States. The processes in use (ordered by size of production) are: (1) oil furnace, (2) gas furnace, (3) thermal, and (4) channel. All the processes except the third, burn hydrocarbons with a paucity of air so that a maximum amount of black (smoke) is produced. The thermal process consists of thermal cracking in batch units by hot bricks.

The pollution common to all the carbon black manufacturing processes is of course the release of carbon particles and amounts to no more than about $0.1\%_w$ of the process weight because effective methods of separating the black from the entraining air are available. The principal losses from the first three processes and at the point of use are attributable to handling and result from such occurrences as leaks in ducts, breakage of bags, poor housekeeping, and the like. No collection system has proved feasible on channel black systems and they still emit large clouds of black smoke. The amount of smoke from the channel process can be roughly indicated by comparing the yields of 2.5 lb of black per 1000 ft^3 of gas in the channel process with the yield of 3-11 lb for the gas furnace. Theoretically, methane should give 31.82 lb on the basis indicated. The thermal process produces up to a 16-lb yield. Oil furnaces also convert about $55\%_w$ of the feed carbon to black, while oil enrichment in gas furnaces and channel processes gives about 2.5 lb of black per gallon. Channel black plants burn gas in a large number of small flames which impinge on channels that are scraped to remove the carbon black.

In addition to the particulate pollution, the low-temperature combustion produces large amounts of carbon monoxide along with hydrogen. The thermal process uses the hydrogen left after the carbon is removed by cracking to heat the furnace for the next batch and to fire boilers for other purposes. Most of the gas used is low in sulfur and therefore causes no sulfur dioxide problem; however, the oils burned may contain up to a few percent sulfur and most of this produces sulfur compounds in the off-gases. The off-gases from the processes, except for the hydrogen of the thermal process, are usually wasted to the atmosphere, but in some countries plants are required to oxidize the carbon monoxide and combustibles before release.

More than one-half of the carbon black manufacture in the United States takes place in Texas and four of the existing five channel black plants are there. An additional one-third of the black is produced in Louisiana.

4. Plastics and Resins

Plastics are often ground into powders which are then molded into desired shapes or made into fibers. Fine powders may escape as dust, or some of the fibers may become airborne. Plastics are made from resins, and resins cause air pollution problems. These problems result from monomer (and low-molecular-weight polymer) vapors and fumes and include odors, allergens, irritants, and surface contaminants.

Toluene diisocyanate (TDI), which escapes during polyurethane manufacture, has received wide publicity and study. It is a strong irritant to the eyes and respiratory tract and has a TLV of 0.02 ppm_v. Methyl methacrylate odor is quite distinctive and not generally considered as highly obnoxious.

The aldehyde, vinyl, styrene, and oily odors emitted by various plastics manufacturing processes are generally the cause of complaints from residents surrounding the plants. Most of the allergenic reactions occur among plant personnel or persons applying the resin coatings. The concentrations of airborne materials diminish rapidly from the point of emission and even TDI causes few problems.

5. Rubber

Raw rubber is prepared for use by variously adding solvents, vulcanizing agents (often sulfur compounds) plus accelerating

or retarding agents (aldehyde-amines, guanidines, and thiuram
sulfides with zinc oxide, stearic acid, magnesium oxides, and
amines or salicylic acid, benzoic acid, and phthalic anhydride),
conditioners (carbon black, zinc oxide, magnesium carbonate,
and clays), plasticizers and softeners (resins, vegetable and
mineral oils, and waxes), antioxidants (alkylated amines), and
pigments and fillers (colors and low-cost materials).

Air pollution problems in the rubber industry involve dusts
from finely ground additive materials, fumes from the mixing
process, oil mists from liquid additives, and odors. A rubber
mill has two heated rollers which turn at different speeds and
shear the mix between them. The introduction of an additive
may produce a cloud of material. Emissions from a mill (except-
ing solvent vapors) average about 0.5 lb/hr. The handling of
dry materials, especially bulk carbon black, is carried out under
negative pressure to prevent dust releases.

6. Soap and Detergent

Soap manufacturing generally produces odor and dust problems which must be solved. Soap is made by treating fats or
oils with sodium or potassium hydroxide, or other alkali. The
fats and oils often come from rendering animal carcasses and
offal and are quite malodorous (60,000 or even 100,000 odor
units; see Chapter 6). After undergoing a series of purification
steps to remove the alkali and objectionable odors, the soap
then enters product preparation to be made into bars, flakes,
or powder.

Detergent manufacture uses petroleum products (kerosene
and benzene) for raw materials. The problems involved in
synthesizing detergents are those associated with refineries
and chemical plants, mostly leakage of chemical materials
(hydrocarbons, acid, chlorine, and so on) from the process.

The detergent is combined with 60-80%$_w$ of builders, fillers, dyes, perfumes, and so on.

Product preparation for soaps and detergents is similar. The product in a paste form is dried on a drum to make flakes or ribbons for bars, or in a spray drier for powders or granules. Spray drying results in a product that must be separated from the gas stream, usually by inertial separators, and a fine dust which passes on through the inertial separators and is removed by filters or scrubbers. These dusts are highly irritating to nasal passages and therefore require high removal efficiency.

7. Pesticides Manufacturing

The pesticides manufacturing industry has developed largely during the past 25 years. This factor and an inherent caution in dealing with poisons have produced relatively few air pollution problems. The production processes give off fine dusts (0.5-10 μm) and organic solvent vapors. Dust problems come from crushing, grinding, blending, and packaging the pesticide. Those pesticides that are liquid are often adsorbed on clays in order to produce a powder that can be applied as a dust. The dusts are carefully controlled, normally by bag filters.

E. Petroleum

The petroleum industry has many air pollutant releases, from the odors emitted by crude oil to the vapors released by tanks that store the finished products for marketing. It is significant that a California refinery is being sued by its neighbors for noise as well as emitted materials. Most air pollutant emissions occur during the refining and the use of the products. Use is covered elsewhere and the discussion here centers on refining.

A petroleum refinery uses fractional distillation to break crude oil into its components, ranging from pentane to asphalt.

TABLE 2-6

Petroleum Refinery Emissions[a]

Process	Dimensions of emission factor	Emission factor
Boilers and process heaters	lb hydrocarbon/1000 bbl oil burned	140
	lb hydrocarbon/1000 ft³ gas burned	0.026
	lb particulate/1000 bbl oil burned	800
	lb particulate/1000 ft³ gas burned	0.02
	lb NO$_2$/1000 bbl oil burned	2,900
	lb NO$_2$/1000 ft³ gas burned	0.23
	lb CO/1000 bbl oil burned	Negative
	lb CO/1000 ft³ gas burned	Negative
	lb HCHO/1000 bbl oil burned	25
	lb HCHO/1000 ft³ gas burned	0.0031
Fluid catalytic units	lb hydrocarbon/1000 bbl of fresh feed	220
	lb particulate/ton of catalyst circulation	0.10[b]
		0.018[c]
	lb NO$_2$/1000 bbl of fresh feed	63
	lb CO/1000 bbl of fresh feed	13,700
	lb HCHO/1000 bbl of fresh feed	19
	lb NH$_3$/1000 bbl of fresh feed	54
Moving-bed catalytic cracking units	lb hydrocarbon/1000 bbl of fresh feed	87
	lb particulate/ton catalyst circulation	0.04[d]
	lb NO$_2$/1000 bbl of fresh feed	5
	lb CO/1000 bbl of fresh feed	3,800
	lb HCHO/1000 bbl of fresh feed	12
	lb NH$_3$/1000 bbl of fresh feed	5

Source	Units	Value
Compressor internal combustion engines	lb hydrocarbon/1000 ft^3 of fuel gas burned	1.2
	lb NO$_2$/1000 ft^3 of fuel gas burned	0.86
	lb CO/1000 ft^3 of fuel gas burned	Negative
	lb HCHO/1000 ft^3 of fuel gas burned	0.11
	lb NH$_3$/1000 ft^3 of fuel gas burned	0.2
Miscellaneous process equipment		
Blowdown system: Controlled	lb hydrocarbon/1000 bbl refinery capacity	5
Uncontrolled		300
Process drains: Controlled	lb hydrocarbon/1000 bbl waste water	8
Uncontrolled		210
Vacuum jets: Controlled	lb hydrocarbon/1000 bbl vacuum distillation capacity	Negative
Uncontrolled		130
Cooling towers	lb hydrocarbon/10^6 gal cooling water capacity	6
Pipeline valves and flanges	lb hydrocarbon/1000 bbl refinery capacity	28
Vessel relief valves	lb hydrocarbon/1000 bbl refinery capacity	11
Pump seals	lb hydrocarbon/1000 bbl refinery capacity	17
Compressor seals	lb hydrocarbon/1000 bbl refinery capacity	5
Others (air blowing, blend changing and sampling)	lb hydrocarbon/1000 bbl refinery capacity	10

[a] From Atmospheric Emissions from Petroleum Refineries--A Guide for Measurement and Control, U.S.P.H.S. Publ. No. 763, NCAPC, Cincinnati, Ohio, 1960.
[b] Without electrostatic precipitator. [c] With electrostatic precipitator.
[d] With high-efficiency centrifugal separator.

The products removed from distillation may be changed into smaller molecules by a catalytic reaction (cat cracking), larger molecules by polymerization or alkylation, or more desirable molecules of approximately the same size by re-forming or isomerization.

The fact that many of the refinery processes are carried out with heat and pressure is responsible for most emissions. A survey made in the Lost Angeles area showed the emissions listed in Table 2-6.

F. Paper

Paper manufacturing has long been notorious for its air pollution, particularly sulfurous odors. The industry is now cognizant of its responsibilities and is proceeding to alleviate the problems.

The kraft paper manufacturing process uses debarked pulpwood as a raw material. Small logs are broken into wood chips which are then put into a digester with aqueous sodium sulfide and sodium hydroxide (white liquor). Cooking for about 3 hr with steam digests the lignin and produces cellulose fibers and a liquid (black liquor). The material enters a blow tank and is separated into solid and liquid fractions.

The pulp enters the papermaking process where it is washed, bleached, pressed, and dried. The black liquor goes to evaporators for concentration. The concentrated black liquor is burned in a recovery furnace to eliminate the organics and leave mostly sodium sulfide and sodium carbonate. The residue is dissolved (green liquor) and treated with lime to convert the sodium carbonate to sodium hydroxide (white liquor). The calcium carbonate is precipitated and burned to give lime.

Table 2-7 shows the particulate and odor sources of the kraft pulping process for papermaking.

TABLE 2-7

Kraft Pulp Emissions

(pounds per ton of dry pulp produced)[a]

Source	Hydrogen sulfide	Methyl mercaptan	Dimethyl sulfide	Particulate pollutants	Type of control
Digester blow system	0.1–0.7	0.9–5.3	0.9–3.8	Negative	Untreated
Smelt tank	--[b]	--	--	20 5 1–2	Uncontrolled Water spray Mesh demister
Lime kiln	1	Negative	Negative	18.7	Scrubber (approximately 80%$_w$ efficient)
Recovery furnace[c]	3.6	5	3	150	Primary stack gas scrubber
	3.6–7.0	--	--	7–16	Electrostatic precipitator
	0.7	--	--	12–25	Venturi scrubber
Multiple effect evaporator	1.2 0–0.5	0.04 0.003–0.030	-- Negative	Negative Negative	Untreated Black liquor oxidation
Oxidation towers	--	--	0.1	Negative	Black liquor oxidation

[a] From P. A. Kenline and J. M. Hales, Air Pollution in the Kraft Pulping Industry, U.S.P.H.S. Publ. No. 999-AP-4, NCAPC, Cincinnati, Ohio, December 1964; A Study of Air Pollution in the Interstate Region of Lewiston, Idaho and Clarkston, Washington, U.S.P.H.S. Publ. No. 999-AP-8, Cincinnati, Ohio, December 1964.

[b] Not available

[c] Gaseous sulfurous emissions are greatly dependent on the oxygen content of the flue gases and furnace operating conditions.

G. Solvents, Paints, and Varnishes

The production and use of solvents, paints, and varnishes
result in emissions into the atmosphere. Solvents generally
evaporate readily and may in themselves be considered air
pollutants. The Los Angeles Air Pollution Control District
(LAAPCD) has been diligent in tracking down and regulating
hydrocarbon sources since their role in Los Angeles smog formation was determined. Cooking varnish and baking paint liberate
several polluting materials.

The most widely used solvents other than water are
chlorinated hydrocarbons. The trend has been away from carbon
tetrachloride and toward tetrachloroethylene or perchlorethylene.
The cost of these solvents often makes recovery attractive, and
releases from degreasing or dry cleaning operations tend to be
small. Dry cleaning accounts for about 4 lb of solvent per
capita-year (< 2 lb chlorinated hydrocarbons and > 2 lb hydrocarbons).

Varnish cooking releases fatty acids, glycerin, acrolein,
phenols, aldehydes, ketones, and terpenes. A glance at the
list indicates that the usual problem involves odors. Paint
application and baking ovens also present odor problems along
with those from particulates and evaporated solvents.

H. Agriculture and Food

Agricultural and food industries cause some of the most
pressing air pollution problems. These problems include dusts,
odors, smokes, and toxic materials from various phases of the
industries.

1. Production

Air pollution from agricultural production runs the gamut
of pollutant types. Dusts come from plowed fields, hay and
grain operations, transport and cattle trails; pollens and

allergenic materials from some crops; odors from feed lots,
manure piles, and barnyards; smoke and combustion products
from clearing land, burning field stubble, and burning refuse;
and toxic materials from the application of pesticides, burning
farm chemical containers, and burning poison ivy.

Undoubtedly the largest amount of agricultural pollution
by weight results from the wind erosion of plowed land in semiarid
regions. Large clouds of dust are moved across parts of
Nebraska, Colorado, Kansas, Oklahoma, and Texas during the
windy spring months. These clouds sometimes attain heights
of more than 20,000 ft and persist for great distances from the
point of erosion. Clouds that originated in West Texas have
been tracked over the Gulf of Mexico several hundred miles
away. Dust storms may become so severe in and near the areas
of origin that visibility is limited to <100 yd at midday.

Odors are particularly severe around feed lots. Beef
animals may be put on a feed lot with a density of one animal
per 15 ft^2. Each animal passes 90-120 lb of fecal matter per
day. It is easy to imagine what odors can develop from the feces
and urine from thousands of beef animals in a small space.
Water pollution problems of feed lots are also severe. Washable
concrete floors can eliminate many of the problems but they are
quite expensive.

Farmers tend to burn wastes if they are at all combustible.
Sometimes large amounts of kerosene or even flamethrowers are
used to destroy green weeds. Smokes from fires that burn poison
ivy, poison oak, or pesticide containers and residues may be
somewhat toxic.

Pesticides have received more attention than any other
agricultural air pollutants. Pesticide application methods result
in considerable amounts of the materials remaining airborne. In
order to distribute pesticide on the undersides of leaves, small

droplets or dusts that diffuse are used. The production of
particulates small enough to diffuse (less than a few microns)
is usually accompanied by the production of many particles
too small to settle out (≤ 1 μm). There is also a tendency to
operate crop sprayers when winds are higher than they should
be and drifting results. Several cases of crop damage from
drifting defoliants have been settled in the courts of Texas
alone.

2. Preparation

The preparation of agricultural products for consumption
often causes dusts, odors, mists, and smoke. Because of the
organic nature of the products, many of these pollutants are
particularly severe.

Dusts are produced in copious amounts by grain milling,
feed grinding and pelletizing, alfalfa and orange pulp drying,
and cotton ginning. Most of these operations use cyclones on
their emission streams; however, as much as 50%$_w$ of the dust
(1-3%$_w$ of process weight) often passes through the cyclone
collector and into the air. Dust explosions have encouraged
grain handlers to keep dust concentrations low.

Coffee roasting gives off particulates and odors. The
particulates come from quenching the hot beans. Odors are
released during the entire process and although the odor is
pleasant in small and occasional amounts, it is quite objectionable on a steady and concentrated basis. The odors are caused
by aldehydes, organic acids, alcohols, and lesser concentrations
of other constituents.

Odor problems are frequently associated with the production of cotton seed oil and meal, soybean oil, corn oil, and so
on. The processes are carried out at elevated temperatures and
acrolein and other odors are emitted.

Cotton gins produce associated dust problems and often smoke problems from the burning of hulls and motes. Both dust and smoke problems are aggravated by the presence of a defoliant applied to the cotton to permit mechanical picking. Arsenic acid is often used as the defoliant. Insurance companies have apparently paid some claims for arsenic poisoning among gin employees, but specific data are not easy to obtain.

Smoke and odor come from smokehouses and barbecues. Odors are also released by frying. Although these sources may not be important for individuals, they are quite significant for commercial establishments.

Meat processing often involves the emission of disagreeable odors. Fecal odors may come from holding pens, blood odor from slaughter floors, wet hair or feather odors from processing rooms, and other odors from by-product processing.

3. Salvaging and By-Products

The normal processing of meats and fish results in considerable amounts of inedible materials--offal. These materials along with dead animals are usually processed for recovery of the marketable fractions. Rendering produces oil and grease for soap manufacture and other uses and protein cake (crackling), which is ground as a supplement to animal feed. Blood is dried and ground for use as fertilizer or feed supplement. Feathers are pressure-cooked, then dried and ground into a high-protein meal.

Dusts are produced in the grinding and handling of meals. Grinding is often done with a hammer mill, and the meal carried by a pneumatic conveyor to a cyclone for separation. The fraction passing the cyclone may be as high as 3 or $4\%_w$ of the process weight.

Small volumes of odorous gases are often released from cookers and driers. These contain amines, aldehydes, organic

acids, and some sulfur compounds. Table 2-8 shows the inten-
sities of odors from some of these processes.

I. Wastewater Treatment

Solving one pollution problem by creating another pollution
problem has received considerably more discussion than remedial
action. Wastewater treatment processes frequently cause air
pollution by air-stripping of volatiles, by injecting particles into
the air through evaporation of wastewater droplets, and by
releasing odious gases from sludge digestion, screenings, and
stable wastewater. Such plants may cause particulate and/or
odor problems by incineration or grinding and bagging of sludge.
Occasionally, froth is blown off aeration tanks and across highways in amounts that cause visibility hazards. The first two
factors occur primarily as a result of aeration for biological
breakdown.

Yang reported more than $80\%_w$ of the benzene in a solution
with an original concentration of 100 mg/liter was air-stripped
in 24 hr. His decay constants (K) for this and other materials
that follow the exponential law are shown in Table 2-9.

$$C = C_0 \exp (-Kt),$$

where C = concentration at time t, C_0 = C for t = 0, and K = air-
stripping decay constant. Air-stripping pollutes primarily with
hydrocarbons and with odors.

Glaser found more than 15,000 particles (>1 μm) per cubic
foot of aeration air from an activated sludge (biosorption) treatment unit. The distribution showed negative skewness with a
mean size of about 12 μm. Randall counted more than 1200
viable bacteria (on nutrient agar) per cubic foot of aeration air
from the aeration tanks of biosorption plant and a conventional
activated sludge plant. About 22% of these bacteria were
identified as enterics, and 6% of the total proved to be known

TABLE 2-8

Odors Released by Rendering and Salvaging[a]

Source	Odor concentration (odor units/scf) Range	Typical	Typical moisture content of feeding stocks (%)	Exhaust products (scf/ton of feed)	Odor emission rate (odor units/ton of feed)[b]
Rendering cooker, dry-batch type[c]	5,000–500,000	50,000	50	20,000	$1,000 \times 10^6$
Blood cooker, dry-batch type[c]	10,000–1 million	100,000	90	38,000	$3,800 \times 10^6$
Feather drier, steamtube[d]	600–25,000	2,000	50	77,000	153×10^6
Blood spray drier[d,e]	600–1,000	800	60	100,000	80×10^6
Grease-drying tank, air blowing 156°F 170°F 225°F		4,500 15,000 60,000	<5	100 scfm per tank	

[a]From J. A. Danielson, Ed., Air Pollution Engineering Manual, U.S.P.H.S. Publ. No. 999-AP-40, NAPCA, Cincinnati, Ohio, 1967.

[b]Assuming 5% moisture in solid products of system.

[c]Noncondensable gases are neglected in determining emission rates.

[d]Exhaust gases are assumed to contain 25% moisture.

[e]Blood handled in spray drier before any appreciable decomposition occurs.

TABLE 2-9

Air-Stripping Constants

Material	K (days^{-1})	C_0 (mg/liter)
Nitrobenzene	0.843	250
Methanol	0.263	1360
Ethanol	0.302	2220
Refinery waste A	0.332	475
Refinery waste B	0.345	782
Benzene	1.71	100
Monochlorobenzene	0.969	100
Aniline	0.198	100

respiratory pathogens--Klebsiella pneumonia and K. oxytocum.
No intestinal pathogens were found although considerable effort
went into testing for shigella and salmonella. The half-life for
the mixed population of bacteria was 0.38 sec; however,
klebsiella should survive longer because of their heavy encapsu-
lation.

J. Earthmoving, Construction, and Demolition

Earthmoving, construction, and demolition generally
cause considerable amounts of dusts and result in release of
smoke and noise as well.

Earthmoving produces airborne dusts in the same manner
as quarrying. The earth being moved and put down is usually
moist to facilitate compaction, but the haul road is often quite
dusty. Earthmoving jobs may be accompanied by smoke from
diesel equipment exhausts and from burning materials cleared
from the site and burning ruined tires. Construction and demolition are usually carried out in haste with little regard to air
pollution. Noise from pneumatic hammers, blasting, and equip-

ment all contribute to the overall din around such projects.
Scrap materials are often burned on site with much smoke emission. These include wood and paper scraps and asphalt shingles
and other combustibles.

The batching of asphalt paving and concrete materials
was described earlier.

Most of the troubles from these industries stem from the
here-today-gone-tomorrow attitudes associated with them.

5. Space Heating and Home Incineration

Space heating and home incineration account for a significant share of the total emissions of tabulated air pollutants
(see Table 2-1). These operations differ from their counterparts
in industry, power plants, and commercial incineration in that
lower combustion temperatures, more emissions, and fewer
emission controls present different problems and are more difficult to control.

The parts of the country that use coal for home fuel have
large amounts of coal smoke (soot), fly ash, and sulfur dioxide
in their atmospheres. Oil-burning regions are usually without
large amounts of particulates but still have high sulfur dioxide
emissions. Those areas that burn natural gas for home heating
have the minimum amount of combustion pollutants. Of course
areas with electric heating are rather free of air pollution from
heating even if the electricity is generated in the area by fossil
fuel. Large plants can afford to clean their stack gases.

Home incineration is objectionable for the same reasons
as heating with coal or other solid fuel, plus the odors of half-
burned garbage and other refuse components. Some cities have
tried to reduce their solid waste problems by permitting or
requiring home incineration. The error of such an approach has
become quite apparent to most of these cities, including New

York. One of the earliest regulations enacted by Los Angeles was that home incineration could not be carried out during periods with little mixing.

In the interests of fire safety, leaves are often burned during the late afternoon when winds have calmed. The calm periods have little mixing and rather severe smoke problems often result when many people burn leaves or grass.

6. Natural

Natural sources dominate in many air pollution problems. The air pollutants derived from nature include swamp gas, salt sprays from the sea; dusts picked up by winds; smoke from naturally set fires; terpenes and resins from forests; pollens from ragweed and forests; fog; noise from thunderstorms; photochemical ozone, nitrogen oxides, and other oxidants; gases, vapors, and particulates from volcanos, geysers, and fissures; and climatic factors (temperature, precipitation, and wind).

Salt spray is air pollution not only along all seacoasts but also at sea where it causes rusting of ships. Droplets of salt water are injected into the air by breaking waves; then the water evaporates leaving salt particles. These particles cause severe corrosion problems (see Chapter 10).

An estimated 30 million tons of dust come from natural sources each year. Much of the dust is in remote areas where no air pollution, within the definition used here, exists. Methane (CH_4) or swamp gas amounts to about 1600 million tons/year and terpenes about 170 million tons/year.

Many forest and prairie fires are set by lightning. The smoke from these fires causes local traffic hazards and drifts over towns and cities, contributing to smoke problems there. Moreover, the combustion products asphyxiate much of the wildlife not killed by the fire.

Fogs (below 25 ft) have been reported to amount to about 15 million tons annually. Fog is a problem in itself and compounds problems from other pollutants.

The amount of pollen put into the atmosphere each year has been reported as 1-2 million tons, but the problems caused by pollen are all out of proportion to the weight (see Chapters 1 and 10).

7. Man-Made Radioactivity

The coming of the nuclear age brought concern for the hazards of ionizing radiation in our air environment--more concern than seems warranted by the probable risks of low-level exposures from worldwide fallout. Radiation emitters, radioactive gases and small particles, are carried by winds. The residence times of radioactive particles and their fallout patterns have provided a valuable tool in tracing circulation patterns of the earth's atmosphere.

Not much radioactivity has been injected into the air by nuclear weapons testing since the moratorium (1963) on nuclear tests by most of the countries of the world. Nuclear explosive devices provide an inexpensive, safe method of excavation; therefore a new Panama Canal and harbors and other surface excavations will likely be done with nuclear explosives as soon as our perspective toward radiation dangers becomes properly adjusted. The Sedan Shot (July 1962) proved the efficacy of such peaceful uses of nuclear explosions.

Most of the man-made radioactivity entering the atmosphere comes from tests made by the French and the Chinese, nuclear reactors, and spent reactor fuel reprocessing--a rather insignificant amount coming from fuel production. One fission explosion may produce many hundreds of megacuries (see Chapter 7) of radioactivity. The decay of this activity varies approximately as a power function with time to the -1.21 or a doubling of the time after the explosion halves the remaining activity.

Argon is activated to ^{41}A in substantial amounts by air-cooled reactors. The Brookhaven National Laboratory reactor puts out thousands of curies per day through a stack. Radioisotopes of oxygen and nitrogen are of no concern except in the immediate vicinity of the reactor because they have half-lives of no more than several seconds.

8. Individual Activities

A person who emits body odors may pollute only a small space around himself, but it is pollution nonetheless. Individual pollutions that cause the most concern to the most people include noisemaking (especially from lawn mowing, hi-fi sets, and fireworks), cigar smoking, outdoor cooking, leaf burning, esthetic fireplace fires, lawn fertilizers, allergenic lawn and garden plants, and other activities that cause odors, noise, and smoke.

9. Spills, Fires, and Other Accidental Releases

Air pollution catastrophes occur from a buildup of pollutants during poor mixing conditions (see Chapter 1) or from accidental releases. Because of the improved communications and vastly increased amounts and varieties of chemicals being handled, people are becoming more aware of the hazards from accidentally released toxic materials.

A train wreck in one of the Plains states released ammonia which killed seven people (February 1969). A nerve gas released by the military during a test at a proving ground in Utah was blamed for the death of thousands of sheep. The nerve gas was supposedly carried across the mountains by the wind, then came to the earth in a rain. Chlorine gas leaked from storage cylinders in Austin, Texas, and forced evacuation of several city blocks. A cloud of acrid fog moved through a low-lying area in Houston after acid truck tanks were washed with too little water.

Warehouse fires involving pesticides and other toxic materials have resulted in exposures to firemen and spectators. Cotton gins have burned motes that contained considerable arsenic from the arsenic acid used as a defoliant.

V. SUMMARY

The delineation of air pollution sources necessarily depends to some extent upon the definition of air pollution. The broad definition that includes noise and heat and cold is used here. Some air pollution sources may result only in local problems or may cause no trouble unless they contribute to the overall levels of emissions from other sources. Widespread heat and cold pollution problems result only from natural processes. Heat islands caused by cities may modify climate slightly but are not of as great significance as pollution problems.

Historically, nearly all pollution has been attributed to the products of combustion and even now this tendency persists to a large degree. Many people do not look upon the products of nature as pollutants, and few inventories of pollutants show dust from wind erosion of plowed fields although people in the southern wheat belt suffer great financial loss and personal distress from these dusts.

Heavy metals industries have been guilty of emitting great plumes of black smoke and sulfur dioxide and in many cases continue to do so. There has been a large measure of control instituted in the Pittsburgh area, but much remains to be done there and in Birmingham and other locations. Many of the particulates from metals manufacture are fumes which are usually quite small (≤ 1 μm) and difficult to control. Sulfur dioxide from roasting pyritic copper and iron ores is often diluted into the atmosphere through tall stacks. This results in regional and local problems.

The Los Angeles smog problem has resulted in focusing
attention on auto exhaust as the principal source of air pollution
in the United States. In most cities auto exhaust is important,
but it is probably not as important as the noise from transporta-
tion--surface and air. There are places in the United States
where neither exhaust nor noise from transportation is significant;
however, the entire country will be controlled in accordance
with the needs of Los Angeles, at least for the immediate future.

Chemical processes often waste significant portions of
their raw materials or products. Sometimes economic recovery
has proved feasible but more often cleanup must be charged to
air pollution prevention.

Airborne releases of radioactivity are quite minimal since
the test ban went into effect. The ability to measure very small
amounts of radioactivity coupled with the belief that little is
known about the effects of radiation causes widespread concern.
Pesticides have received much attention for the same reasons.
The benefits are seldom mentioned by those who condemn pesticides.

PROBLEMS

1. Is an air cleaner capable of removing 3-μm particles likely
to be effective on wet process cement kiln dust? Why?

2. What effect on the total inventory of air pollution emissions
would be the classification of carbon dioxide as a pollutant?

3. Suggest units for presenting the amount of noise pollution
(since weight is not applicable).

4. What pollutions other than noise are not amenable to weight
categorization?

5. What other pollutants could be added to the inventory of
Table 2-1?

6. What are some sources of air pollution not covered in this
chapter?

BIBLIOGRAPHY

Atmospheric Emissions from Petroleum Refineries: A Guide for
    Measurement and Control, U.S.P.H.S. Publ. No. 763,
    Washington, D. C., 1960.

Atmospheric Emissions from Nitric Acid Manufacturing Processes,
    U.S.P.H.S. Publ. No. 999-AP-27, Cincinnati, Ohio,
    1966.

Atmospheric Emissions from Sulfuric Acid Manufacturing Proc-
    esses, U.S.P.H.S. Publ. No. 999-AP-13, Cincinnati,
    Ohio, 1965.

J. A. Danielson, Air Pollution Engineering Manual, U.S.P.H.S.
    Publ. No. 999-AP-40, Cincinnati, Ohio, 1967.

R. L. Duprey, Compilation of Air Pollutant Emission Factors,
    U.S.P.H.S. Publ. No. 999-AP-42, Durham, North
    Carolina, 1968.

G. R. Fryling, Ed., Combustion Engineering, Rev. ed.,
    Combustion Engineering, Inc., New York, 1966.

J. R. Glaser and J. O. Ledbetter, "Sizes and Numbers of Aer-
    osols Generated by Activated Sludge Aeration," Water
    & Sewage Works, 114:6, 219-221, June 1967.

L. R. Hafstad, "Automobiles and Air Pollution," Universities,
    National Laboratories, and Man's Environment, Proceedings
    of Conference July 27-29, 1969, Chicago (USAEC, Divi-
    sion Technical Information, November 69), pp. 109-132.

T. E. Kreichelt, D. A. Kemnitz, and S. T. Cuffe, Atmospheric
    Emissions from the Manufacture of Portland Cement,
    U.S.P.H.S. Publ. No. 999-AP-17, Cincinnati, Ohio, 1967.

Motor Vehicles, Air Pollution, and Health, U.S.P.H.S., U.S.
    Government Printing Office, Washington, D. C., 1962.

J. O. Powers, "Airborne Transportation Noise: Its Origin and
    Abatement," Sound Vibration, 2:6, 10-17, June 1968.

Process Flow Sheets and Air Pollution Controls, American
    Conference of Governmental Industrial Hygienists,
    Cincinnati, Ohio, 1961.

C. W. Randall and J. O. Ledbetter, "Bacterial Air Pollution from
    Activated Sludge Units," Am. Industrial Hyg. Assoc.
    J., 27:6, 506-519, November-December 1966.

W. H. Rupp, "Air Pollution Sources and Their Control," Air Pollution Handbook, Section 1, McGraw-Hill, New York, 1956.

A. C. Stern, Ed., Air Pollution, Vol. I, 1st ed., Academic Press, New York, 1964.

A. C. Stern, Ed., Air Pollution, Vol. III, 2nd ed., Academic Press, New York, 1968.

W. S. Smith, Atmospheric Emissions from Fuel Oil Combustion-- An Inventory Guide, U.S.P.H.S. Publ. No. 999-AP-2, Cincinnati, Ohio, 1962.

W. S. Smith and C. W. Gruber, Atmospheric Emissions from Coal Combustion--An Inventory Guide, U.S.P.H.S. Publ. No. 999-AP-24, Cincinnati, Ohio, 1966.

G. J. Thiessen and N. Olson, "Community Noise--Surface Transportation," Sound Vibration, 2:4, 10-16, April 1968.

J. W. Vogh, "Nature of Odor Components in Diesel Exhaust," presented at the Air Pollution Control Association Meeting, St. Paul, Minnesota, 1968.

J. T. Yang, Development of Design Parameters for Biological Treatment of Industrial Wastes, Dissertation, Univ. of Texas, Austin, June 1968.

Chapter 3

ATMOSPHERIC TRANSPORT OF POLLUTANTS

I. INTRODUCTION

Gases, and particles small enough to remain airborne for significant periods, may be carried long distances by the movements of the air. As a result of such carriage, the pollutants can contribute not only to local air pollution problems but also to regional difficulties; however, dispersion in the air often dilutes pollution to prevent noxious concentrations of pollutants from reaching the ground (see Chapter 15).

Long-distance movements result from winds and from the transverse horizontal and vertical spreads from turbulence. All major air pollution catastrophes have occurred during periods of low winds and limited vertical and horizontal mixing.

II. WINDS

Winds are caused by pressure differences, which in turn result from temperature differences. Winds may occur over a large area and be caused by large-scale climatic conditions (e.g., the westerlies, the jet stream, frontal movements, cyclones--lows, or anticyclones--and highs), or over a small area and be caused by local geographical and topographical features (e.g., sea breeze and upslope and downslope winds). The local wind direction can be influenced by channeling along valleys, eddy circulations behind abrupt drops, or friction with the ground, trees, buildings, and so on.

A. Geostrophic Winds

Logic dictates that wind direction be along a decreasing gradient (perpendicular to the isobars, lines of equal pressure); however, this is not the case. The pressure gradient force is opposed by the coriolis force, a force from the rotation of the earth (the same force that causes a sink vortex to be counter-clockwise in the Northern Hemisphere), and wind direction is more or less parallel to the isobars. When the coriolis force equals the pressure force, which is practically the situation above surface friction (>~3000 ft), geostrophic winds result. These winds move in a direction parallel to the isobars and to the right (Northern Hemisphere) as the observer faces the direction of decreasing pressure (see Fig. 3-1a).

B. Surface Winds

Nearer the surface of the earth an additional force, surface friction, is active. Surface friction swings the winds back toward the low pressure (see Fig. 3-1b). The resultant of the coriolis and friction forces is equal to and opposite the pressure force.

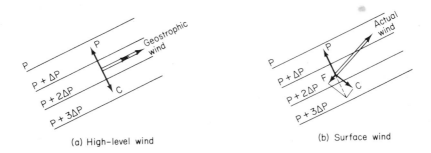

(a) High-level wind

(b) Surface wind

FIG. 3-1. Winds and pressure fields. (a) High-level wind. (b) Surface wind.

The force equations for low-level winds (assuming a flat pressure field) are

$$du/dt - 2v\Omega \sin \Phi = - 1/\rho \; \partial p/\partial x + F_x ,\qquad(3-1)$$
$$dv/dt - 2u\Omega \sin \Phi = - 1/\rho \; \partial p/\partial y + F_y ,$$

and

$$dw/dt - 2u\Omega \sin \Phi = - 1/\rho \; \partial p/\partial z - g,$$

where u, v, w = x, y, and z components of wind velocity, Ω = angular velocity of the earth, Φ = latitude, ρ = density, p = pressure, F_i = ith component of friction force, and g = acceleration of gravity. The terms containing $\sin \Phi$ represent the coriolis force.

Wind forces also have a centrifugal component in actual circulations, a force which is added to the P value for highs and is subtracted for lows; therefore higher wind velocities result from the same pressure gradient in highs as in lows.

Since pressure (density) is related to temperature, the direction of the wind is also related to the temperature gradient. The difference vector between winds at two different levels is parallel to the isotherms (lines of equal temperature).

C. Systematic Variations of Local Surface Winds

The velocity of the wind commonly increases with height, and surface wind decreases at night under usual conditions. Wind velocity increases with height because of decreasing surface friction. There are elaborate formulas for calculating this variation; however, the most used and one that adequately serves many purposes is the power law or

$$u/u_0 = (z/z_0)^\alpha ,\qquad(3-2)$$

where u = wind speed at height z, u_0 = wind speed at height z_0, and α = parameter of fit $\approx 1/7$ for 2- to 10-m heights. The power law fits actual daytime situations rather well, but the nighttime regime is quite different (see Fig. 3-2). Figure 3-2 also shows

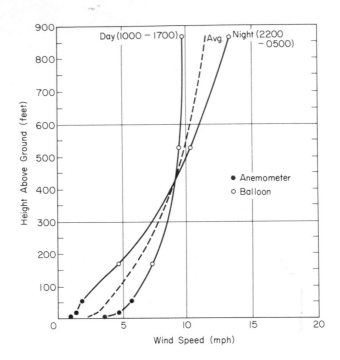

FIG. 3-2. Wind variation with height. From J. Z. Holland, Meteorological Survey of Oak Ridge Area, U.S. Atomic Energy Commission Rept. ORO-99, Oak Ridge, Tennessee, 1953.

that surface wind speed is less at night, while that aloft increases. This fact agrees with the time of breakup of the nocturnal inversion.

D. Cyclones and Anticyclones

Cyclones and anticyclones are large-scale circulations which are often several hundred miles across. Cyclones (lows) cause counterclockwise circulations in the Northern Hemisphere and highs cause clockwise circulations. Most of the weather effects are associated with lows; air pollution problems occur more with highs.

E. Frontal Movements

Large masses of air moved by circulation are often different in temperature from the region toward which the air is moving. The juncture of such a mass with the local air is called a front, warm or cold as it relates to the local temperature ahead of it.

F. Geographical Effects on Winds

Geographical or topographical features often determine local winds and their directions. The sea breeze and valley winds are the most important topographic winds in air pollution analysis.

The sea breeze is caused by the more rapid heating of land than water. This heating makes air rise over land and brings in heavier air from over the water. The reverse, a land breeze, occurs on the cooling cycle but is less pronounced than the sea breeze.

Valley circulations are caused by heating and cooling. The winds blow upslope (perpendicular to the axis of the valley) in the forenoon and downslope in the late evening. Along the axis of the valley, the winds blow upvalley in the late afternoon and downvalley in the early morning or late at night. Valleys can also cause channeling of cross winds that strike the valley obliquely and change their direction so that they travel along the axis of the valley.

G. Measurement of Winds

It is important to the analysis of an air pollution problem to know the magnitudes both of the transporting mechanisms, winds and turbulence, and of the factors that cause those forces, pressure and temperature. The humidity and cloud cover or insolation also must be known, for they influence the lifetimes of airborne pollutants.

Winds are usually measured only horizontally for direction
and velocity. An object or material carried by the wind may be
released and its position determined at various times after release, or a stationary instrument may measure the parameters at
a given location.

1. Measurement by Carriage

Most commonly, the object released for wind measurement
is a balloon. The types of balloons used are the pibal (pilot
balloon), the Rawinsonde (radiosonde), and the tetroon. The
pibal is inflated to a given buoyancy and released. Its position
is determined by measuring the angles of elevation and azimuth
with a theodolite and assuming a rate of rise, or by two separately positioned theodolites without regard to the rate of rise.
The change in position between sightings indicates the direction
of the wind at that height, and the distance moved in the time
interval determines the velocity.

The Rawinsonde balloon is filled and released in the same
manner as the pibal but is traced by a radio direction finder or
by radar. The balloon carries a small transmitting apparatus
which sends pressure, temperature, and humidity data to the
ground station.

The rates of rise of the pibal and the radiosonde balloons
are about 500 and 1000 ft/min, respectively. This means that
positions must be determined at 30- or 15-sec intervals and even
then most details of the near-ground wind structure are lost.
The famous Brookhaven National Laboratory photograph (see Fig.
3-3) showing three entirely different wind directions at heights
of 50, 150, and 350 ft proves the importance of detail in wind
structure.

The tetroon is a tetrahedral balloon which is inflated sufficiently to give it the same buoyancy as the surrounding air,

FIG. 3-3. Variation of wind direction. From M. E. Smith, "Reduction of Ambient Air Concentrations of Pollutants by Dispersion from High Stacks," Proc. 3rd Natl. Conf. Air Pollution, Washington, D. C., December 12-14, 1966.

then released at a given height and tracked. Close-in tracking
may be done as for the pibal; however, long-distance tracking is
best accomplished by helicopter but is sometimes done with an
airplane or an automobile or radar. The most desirable situation
is to release several tetroons of various colors at different heights
simultaneously.

A puff of visible material (methylene blue) or a material that
can be found readily by analysis (SF_6) may also be used. More
often a continuously emitted plume (comparable to that from a
smokestack) or a streak of smoke left by a rocket is used. The
number of positions requiring monitoring discourages analytical
measurements of a gas release except in unusual cases, and the
normal practice is to employ a visible material. Sometimes a
fluorescent dye material with small particle size is released and
sampled for with rotorod impactors. A rocket trail is observed
or photographed at various synchronized times by two observers
(or cameras) watching the trail at right angles to each other.

The balloon or material release method is expensive and
sometimes objectionable, particularly near airways; therefore
wind measurement by this method is not made on a wide scale.

2. Measurement by Fixed Instrumentation

The method commonly employed in wind measurements is
to set velocity and direction instruments at a point and measure
the winds passing that location. Such instruments are frequently
placed on a building top or on a tower about 10 m off the ground.
Sometimes a kytoon (captive balloon) is used to position instruments at a height where no tower is readily available. Measurements are made at some height to avoid influences from the
ground.

Wind direction is determined by a vane that is free to rotate in a horizontal plane and is designed to align itself with the
wind. The azimuth angle of the vane can be measured by a

sliding-contact rheostat (changing resistance), or by an optical
system. The azimuth given for a wind is the direction from which
the wind comes. This must be kept firmly in mind for transport
determinations in which the interest is in the direction that the
material is carried. The ideal vane would have no size and no
inertia, but since this is impossible the compromise is to develop
a vane that reacts rather swiftly yet does not overreact to a wind
direction change. These specifications have led to the design
of aerodynamic and triangular-shaped vanes rather than flat ones
for general purposes. Sometimes instruments with large damping
(slow response) are used so that the indication is an average of
some fluctuations.

Mounted on the wind vane or nearby is a velocity-measuring
instrument. This instrument may be a propeller which is kept
pointing into the wind by the vane, or a cup anemometer (see
Fig. 3-4). An anemometer measures wind speed by the revolution
rate (angular velocity) of the rotor. The determination is usually
made electrically, either by a contact which gives a pulse, the
rate of which is measured, or by a generator turned by the rotor,
which gives a voltage that is a function of revolutions per minute.

A mechanical signal can also be used for wind speed, a
deflecting vane or a pitot tube-manometer arrangement. A hot-
wire anemometer (costing about $250 to $400) is often used as
a field survey instrument. A heated wire, which changes resistance with temperature and changes temperature with wind
speed, is included in a Wheatstone bridge circuit. A hot-wire
anemometer often measures temperature as well as wind speed
but is self-correcting for temperature when wind speed is being
measured.

Signals from wind vanes and anemometers may be read at
intervals by an observer but are usually recorded on strip charts.
A dual-channel recorder which gives both traces on a single

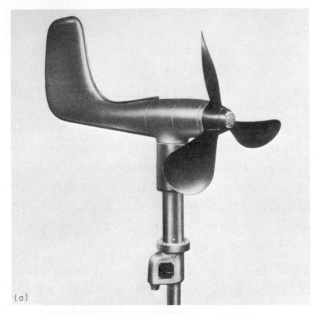

FIG. 3-4. Transmitting wind instruments. (a) Bendix-Friez (propeller-vane) anemometer. (b) Gill cup anemometer and vane. (c) Gill anemometer bivane.

(c)

piece of paper is best suited for this purpose. The traces make
the determination of the average wind speed and direction and
the standard deviations from the averages relatively simple to
obtain (see Chapter 4, Section XIV).

A bivane anemometer (see Fig. 3-4c) measures the azimuth
and elevation angles and the wind speed in the direction it is
blowing. Analysis of bivane data is feasible with computers.

H. Wind Rose

Wind data at a given point are usually represented by a
wind rose. A wind rose is a plot of direction and velocity of the
winds which shows what fraction of the time a given direction and
velocity prevail (see Fig. 3-5). In the plot, up is taken as north
and right as east. The spike plotted in a given direction means
that the wind is blowing from that direction. The width or color

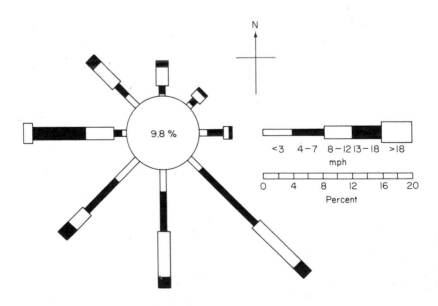

FIG. 3-5. Wind rose.

or hatching of the spike indicates wind velocity, and the length
is proportional to the percent of time a given velocity and direction were observed.

An eight-spike wind rose is detailed enough for air pollution
analyses and is as detailed as wind data usually warrant. Six-
teen points often require bias removal and considerably increase
the amount of work involved in using the wind rose. The principal fault in the wind rose is that wind speeds too low to register
on an anemometer or to move the vane are shown as calms in the
center of the plot. The low wind speeds can be very important
to air pollution applications as was learned by a major city that
located its industrial park so that prevailing winds would carry
pollution away from the city but the prevailing low-speed winds
(calms) were in the opposite direction and frequently brought
stifling pollution into the city. The direction should be reported
whenever known, and anemometers with low starting speeds and
sensitive wind vanes should be used to avoid missing significant
wind data. The calms are often distributed into the wind rose
spikes in proportion to the frequency of the light (0-7 mph) winds
measured.

I. Role of Winds in Air Pollution Analysis

Winds carry air pollution; they also dilute it. The most
serious air pollution conditions nearly always result during periods
of very low winds. The importance of wind dilution may be
readily inferred by considering that up to several hundred miles
of wind (summation of wind speeds times the term of persistence)
may pass a point in 1 day.

Stagnating conditions usually occur with anticyclones,
despite the stronger winds for the same pressure gradient in
highs as in lows. Pressure gradients are often less in highs
than in lows. Figure 3-6a shows the number of stagnation periods that lasted for four or more days in the eastern part of the

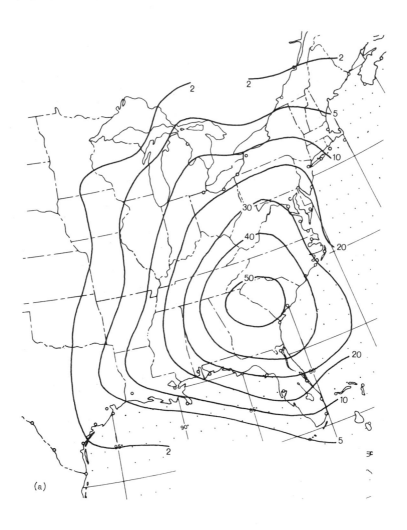

FIG. 3-6a. Map of periods with stagnating conditions.
From J. Korshover, "Synoptic Climatology of Stagnating Anticyclones," Tech. Rept. A60-7, U.S.P.H.S., Cincinnati, Ohio,
1960. (Furnished by Texas State Climatologist)

FIG. 3-6b. Occurrence of periods suitable for stagnation. From G. C. Holzworth, "A Study of Air Pollution Potential for the Western United States," *J. Appl. Meteorol.*, **1**:3, 366-382, September 1962.

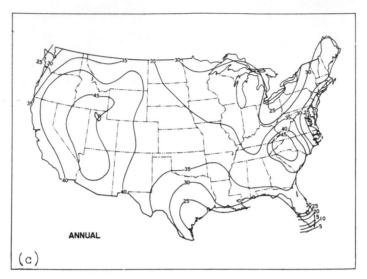

FIG. 3-6c. Inversion occurrence (percent total hours).
From C. R. Hosler, "Low-Level Inversion Frequency in the
Contiguous United States," Monthly Weather Rev., 89, 319-
339, September 1961.

United States during the period 1936 to 1956 as reported by Korshover. October had the highest incidence of such episodes with
only 20% of the total during November through March. Holzworth
found that the 1949-1956 data for the western United States indicated 90% of the semistationary anticyclones shown by numbers
in Fig. 3-6b occurred in October through February. The total
inversion time presented in Fig. 3-6c is of little concern because
the short periods (diurnal) make up a large portion of the total.

Valley winds are of major importance to air pollution study
because many industrial pollution sources require cooling water
for their operations and are therefore located beside a stream or
a lake in a valley. Without winds the valley simply fills up with
the emitted pollutants.

III. CONVECTIVE CURRENTS

Winds are related to temperatures as mentioned earlier.
Vertical movements of the air are particularly dependent upon the
local vertical temperature structure.

A. Lapse Rate versus Stability

A clear understanding of buoyant (thermal) stability is requisite to any analysis of the transport of pollution by the atmosphere. The stability is determined by comparing the lapse rate (temperature change with height, usually in a negative direction, hence the word lapse) with the dry adiabatic lapse rate (about $5.4°F/1000$ ft or $9.6°C/km$, or roughly $1°C/100m$). The dry adiabatic lapse rate is the temperature change that a dry parcel of air would undergo if moved upward adiabatically (without exchange of heat with the surroundings). For our purposes, the dry adiabatic lapse rate applies sufficiently well to moist air as long as the air is unsaturated.

If the lapse rate is superadiabatic (see Fig. 3-7a), a parcel of unsaturated air moved up along the dry adiabatic will be warmer than its surroundings and will continue to rise. Likewise, a parcel moved downward will be cooler than its surroundings and will continue to fall. This condition favors mixing and is termed unstable.

Similar logic applied to subadiabatic conditions leads to the conclusion that a parcel moved up (or down) is cooler (or

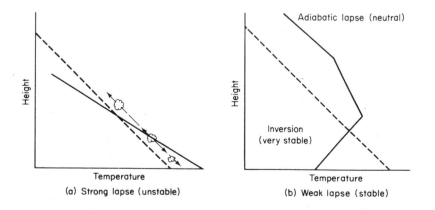

FIG. 3-7. Buoyant stability with lapse rate.

warmer) than its surroundings and returns to the place of origination. Such a lapse tends to prevent mixing and is termed stable.

The intermediate lapse rate, the same as the dry adiabatic, is a neutral stability condition because it neither favors nor disfavors mixing. A parcel of air moved under the neutral state remains at its new position.

The movements that initiate the mixing referred to here are caused by mechanical movements as air is passed over obstacles by the wind.

B. Inversions

At times the temperature of a layer of air actually rises with height over some distance and that layer is called an inversion (see Fig. 3-7b).

Nocturnal inversions occur because the ground cools faster than the air above it. The nocturnal inversion is often intense (10-25°F rise) over a layer from the ground up to a few or several hundred feet. These shallow inversions normally break up from thermal mixing by 10 or 11 a.m.

When winds cause mixing of a stable atmosphere, the resultant lapse is dry adiabatic below the mixing level, with an inversion formed at the top of the mixing. Subsidence inversions occur if a cool layer of air descends below a warmer layer because of different densities. Subsidence inversions cause most of the Los Angeles inversion problems. Warm air moving across cold water forms an inversion near the surface that persists for some time as the air moves inland.

C. Maximum Mixing Depth

The maximum mixing depth (mmd) is the height to which mixing is expected during the day because of the warmup (see Fig. 3-8). The concept is important in the air pollution potential forecasts and average values for an area help to plan certain

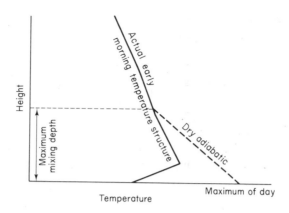

FIG. 3-8. Maximum mixing depth (mmd).

activities that depend upon atmospheric dilution. The mmd is based on the assumption that only vertical temperature variations control vertical mixing. The smallest values of this parameter, and therefore the most troublesome, occur from October through February. Holzworth found the mean mmd values less than 1500 m over the entire United States, except southern Florida during the months cited. Normally, the lowest values are those in December and January, which often average less than 500 m for the month.

IV. TURBULENCE

Atmospheric turbulence determines to a large extent what concentrations of pollutants prevail at points downwind from the emission. Turbulence in air flow results from eddy circulation cells caused by thermal or mechanical forces and are superimposed on the average wind vector. The eddy is visualized as a closed circulation and may range in size from several molecules to more than a mile across.

Measured wind speeds usually fluctuate about some mean almost randomly with time. Consider the passage of an eddy cell;

if a decrease in azimuth is noted at a point, the opposite side of
the cell will cause an increase in azimuth when it passes the
wind vane. The largest changes in direction occur when the
middle of the cell passes over the instrument, while the largest
changes in wind magnitude take place when the eddy cell becomes
tangent to the point of instrumentation. Fluctuations in wind
velocity are termed gustiness. Gustiness is normally used as
the measure of turbulence intensity. Where u, v, and w are the
orthogonal x, y, and z components, respectively, of the wind,
u', v', and w' are used to represent instantaneous component
fluctuations from the mean. Gustiness in a given direction is
defined by

$$g_x = \sqrt{(\overline{u'^2}/\overline{u^2})}, \; g_y = \sqrt{(\overline{v'^2}/\overline{u^2})}, \qquad\qquad (3\text{-}3)$$

or

$$g_z = \sqrt{(\overline{w'^2}/\overline{u^2})}.$$

A parameter that is much easier to measure but is related to the
intensity of turbulence (for a wind speed) is the standard deviation of the azimuth angle. The use of this factor is discussed
subsequently in Section V, B, 5.

In addition to intensity, another factor is needed to characterize turbulence--scale. Scale is the size of the eddies, usually taken as the cell diameter. An approximate measure of scale
is made by taking the period of 1 cycle of change, multiplying by
wind speed, and dividing by 0.67. The last factor is derived
from probability considerations that the expected length of a
random chord passed through a sphere will be 0.67 diameters.
The length of the period is from the time that an azimuthal trace
crosses the average trace line until the next time it crosses
moving in the same direction toward either the smaller or larger
angle. The average period is usually employed and that is merely
the elapsed time divided by half the total number of peaks or the
number of peaks above (below) the average line.

V. DISPERSION MODELS

Historically, atmospheric dispersion models were based on Fick's molecular diffusion equation and later on statistical theory. Recent models have retained the statistical nature of the process but have become much more empirical with the availability of more good data.

A. Instantaneous Point Sources

A source that starts essentially as a point source, or a puff that can be extrapolated back to point source conditions, in the absence of currents is diffused by molecular forces. Fick's equation for molecular diffusion is

$$\nabla^2 X = 1/\not{D} \; \partial X/\partial t, \tag{3-4}$$

where X = concentration diffusing substance, $\not{D}$ = molecular diffusivity, and t = time.

While Fick's equation was never intended for dispersion by eddy diffusion, in which molecular diffusion is so slow as to be ignored without serious error, it was used as is for isotropic turbulence and adapted for nonisotropic turbulence to

$$\partial X/\partial t = \partial (K_x \partial X/\partial x)/\partial x + \partial (K_y \partial X/\partial y)/\partial y + \partial (K_z \partial X/\partial z)/\partial z, \tag{3-5}$$

where K_i = diffusion coefficient in the ith direction. This equation allowed different diffusivities for each direction but still ignored the statistical nature of concentration variations across the puff or cloud.

Sutton's equation is based on the statistical concept of the eddy diffusion process; the equation is

$$X = \frac{M}{\sqrt{\pi}^3 C_x C_y C_z (ut)^{3(2-n)/2}} \exp\left[\frac{-1}{(ut)^{2-n}}\left(\frac{x^2}{C_x^2} + \frac{y^2}{C_y^2} + \frac{z^2}{C_z^2}\right)\right] \tag{3-6}$$

where X = concentration of emitted material at x, y, z, and t; M = mass of material released at $t = 0$, $x = y = z = 0$; C_i = virtual diffusion coefficient in the ith direction (see Table 3-1),

TABLE 3-1

Sutton's Generalized Eddy Diffusion Coefficients[a]

Height of source above ground (ft)	Large lapse, n = 0.20		Zero or small lapse n = 0.25		Moderate inversion, n = 0.33		Large inversion, n = 0.50	
	C_y	C_z	C_y	C_z	C_y	C_z	C_y	C_z
0	0.42	0.24	0.24	0.14	0.15	0.09	0.12	0.07
33	0.42	0.24	0.24	0.14	0.15	0.09	0.12	0.07
82		0.24		0.14		0.090		0.070
100		0.23		0.13		0.085		0.065
150		0.21		0.12		0.075		0.060
200		0.19		0.11		0.070		0.055
250		0.18		0.10		0.065		0.050
300		0.16		0.09		0.055		0.045
350		0.13		0.07		0.045		0.035

[a]From O. G. Sutton, "The Problem of Diffusion in the Lower Atmosphere," Quart. J. Royal Met. Soc., 73, 257, 1947.

[b]C has dimensions $ft^{n/2}$.

u = average wind velocity in x direction, t = time since release, n = dispersion parameter (see Table 3-1), x = horizontal distance from center of cloud in direction of u, y = horizontal distance from center of cloud in direction perpendicular to u, and z = vertical distance from center of cloud in direction perpendicular to x and y. The origin moves with the cloud in all of the instantaneous point source calculations. For calculating ground concentrations, the X value above is doubled to account for ground reflection. If the material is released with dispersions too large to be assumed a point, a virtual origin can be estimated. Some observers use a diameter of one-fifth the distance to the place of interest as the limiting size without correction of travel distance. Holland advocated the use of

$$x_0 = \left[\frac{2Q}{\pi^{3/2} C^3 X_0} \right]^{2/[3(2-n)]} , \tag{3-7}$$

where X_0 = central concentration at $t = 0$. The value of x_0 is then added to the x values in Sutton's equation, including the ut form. The parameters of Sutton's equation are related to the winds as

$$n/(2-n) = \alpha , \tag{3-8}$$

where α = vertical wind structure parameter of Eq. 3-1 and

$$C_i^2 = \frac{4\nu^n}{(1-n)(2-n)u^n} g_i^{2(1-n)} , \tag{3-9}$$

where ν = kinematic viscosity of air ≈ 0.148 at $20^\circ C$ and 1 atm.

EXAMPLE 3-1: If 300 m^3 of sulfur dioxide is suddenly released from an industrial plant at 30 m above the ground into 2 m/sec wind during a strong inversion, what will the concentration be at the center of the cloud 200 m downwind? At the ground?

$C_x = C_y = C_z = 0.065$ $\quad$ $n = 0.50$ (from Table 3-1)

For the cloud center: $\quad X_0 = 64/0.024 = 2670$ gm/m^3

$\quad$ Assuming $20^\circ C$ $\quad M = (64$ gm/gmw$)/0.024$ m^3/gmw

$\quad C^3 = 2.75 \times 10^{-4}$ $\quad X\ 300$ m$^3 = 8 \times 10^5$ gm

$$ut = x + x_0 \qquad x_0 = \left[\frac{2(8 \times 10^5)/2670}{5.63\ (2.75 \times 10^{-4})}\right]^{2/[3(2-0.5)]}$$
$$= 297 \text{ m}$$

$$X = \frac{8 \times 10^5}{5.63\ (2.75 \times 10^{-4})\ (200 + 297)^{3(2-0.50)/2}} \exp\ (0)$$
$$= 0.1064 \text{ gm/m}^3 = 106.4 \text{ mg/m}^3$$
$$= 0.1064 \times 10^6/2670 \text{ ppm}_v = 39.9 \text{ ppm}_v$$

For the ground:

$$X = 2\ (0.1064)\ \exp\left[-1/497^{2-0.50}\ 30^2/0.065^2\right]$$
$$= 0.2128 \exp\ (-19.36) = 0.2128/256 \times 10^6$$
$$= 8.33 \times 10^{-10} \text{ gm/m}^3 = 8.33 \times 10^{-4} \text{ }\mu\text{g/m}^3$$

Other assumptions lead to different solutions for the example. Another method of estimating x_0 would be to guess the angle of spread at, for example, 3° under a strong inversion (see Fig. 3-13), then divide the 8.3-m diameter for a 300-m^3 sphere by the tangent of 3° to obtain a value of 159 instead of the 297 indicated by Holland's formula.

B. Continuous Point Sources

Continuous point source geometry applies to most air pollution emissions. Continuous emission by a point source results in a plume which is carried by the wind and spread by eddy diffusion. The amount of dilution and the height of rise of the plume depend upon the intensity and scale of the turbulent eddies, the wind speed, and the buoyant stability.

1. Plume Types

Plumes released by point sources assume practically an infinite variety of shapes and sizes as they are acted upon by the different wind combinations. Church classified the plumes to be expected with various stability conditions (see Fig. 3-9). It is useful to keep this figure in mind when estimating ground concentrations from plumes.

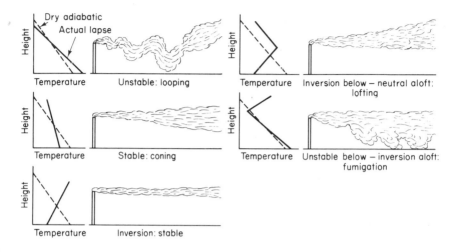

FIG. 3-9. Classification of plume types (after Church).

2. Stack Height

Continuous point source emissions are normally from a stack, sometimes from a duct end. The height of the stack and the height of rise of the plume above the stack play a major role in the ground concentration expected downwind. The actual stack height (H) is, of course, easy to determine. The effective stack height (H_e) is the total height to which the plume rises (see Fig. 3-10) and is quite difficult to estimate with any degree of accuracy.

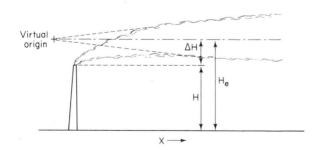

FIG. 3-10. Plume rise.

The rise (ΔH) of the plume above the top of the stack results from two factors, jetting and thermal buoyancy. A rule of thumb is $2\frac{1}{2}$ ft of ΔH per degree Fahrenheit of excess heat (stack temperature minus ambient). There has been a real paucity of plume rise data but not of theories and equations to fit these data. Only five of these many equations are discussed here.

a. Holland (Oak Ridge) Formula. Holland fitted 137 observations from three stacks with heights of 160, 180, and 200 ft. His formula for average conditions is

$$\Delta H = \frac{1.5v_s d + 3 \times 10^{-4} Q_h}{u} \text{ ,} \tag{3-10}$$

where ΔH = rise of plume above the stack (ft), v_s = exit velocity of stack gas (mph), d = stack diameter at exit (ft), Q_h = excess heat in stack gas (cal/sec) = $Q_m c_p (T_s - T_a)$, Q_m = mass flow rate of stack gas (gm/sec), c_p = specific heat of stack gas at constant pressure (cal/gm-$^\circ$C) ≈ 0.24 for air, T_s = temperature of stack gas ($^\circ$C), and T_a = ambient temperature ($^\circ$C). The temperatures, velocities, and stack diameter can be measured rather easily. The stack exit velocity (for air) can also be calculated from

$$v_s = \frac{0.962 \, Q_m T_s}{d^2 P} = \frac{0.032 \, Q_m T_s}{d^2} \tag{3-11}$$

where v_s = stack draft velocity (fps), Q_m = mass emission rate (lb/sec), P = atmospheric pressure (in. of Hg) ≈ 30 in., d = diameter of stack (ft), and T_s = temperature of stack gas ($^\circ$R). Thomas modified Eq. 3-10 by doubling the coefficient of the excess heat term (to 6×10^{-4}). His observations further indicated that the rise calculated by this modified Holland formula should be increased 10-25% for unstable conditions and decreased a like amount for stable conditions, as was recommended by Holland

b. Carson and Moses Equation. One recent study of plume rise by the TVA showed that the Carson and Moses equation gave the best fit to field observations. This equation is

$$\Delta H = A \frac{4.12\, v_s d + 9.22/Q_h}{u} \qquad (3-12)$$

where terms are the same as above except, A = stability correction = 1 for $d\Theta/dz \geq -0.22^\circ C/100$ m = 2.43 for $d\Theta/dz < -0.22^\circ C$ 100 m, $d\Theta/dz$ = change in potential temperature, and ΔH is in meters, v_s is in meters per second, d is in meters, Q_h is in 10^4 cal/sec, and u is in meters per second.

c. <u>Bryant-Davidson Formula</u>. This formula, which was derived from wind tunnel experiments, is generally believed to underestimate the plume rise. The formula is

$$\Delta H = d\, (v_s/u)^{1.4} [1 + (T_s - T_a)/T_s], \qquad (3-13)$$

where all terms are as defined for Eq. 3-10, but the units of T_a and T_s were said to be in degrees Kelvin. ΔH has the same units as d. Here again stability must be taken into account separately since no provision for such is made in the formula.

d. <u>CONCAWE Equation</u>. The most recent information from the TVA study (Thomas et al.) is that the CONCAWE equation gives the best results. It is

$$\Delta H = 0.175\, Q_h^{1/2}/u^{3/4}, \qquad (3-14)$$

where ΔH is in meters, Q_h is in calories per second, and u is in meters per second.

e. <u>Bosanquet et al. Formula</u>. The Bosanquet et al. formula has some internal recognition of atmospheric stability. It has the potential temperature gradient incorporated as a measure of the stability; potential temperature is the temperature a parcel of dry air would have if brought adiabatically from its pressure position to 1000 mbars. The equation, which is probably unduly elaborate, is

$$\Delta H = \frac{4.77}{1 + 0.43\, u/v_s} \sqrt{(Qv_s/u)} + 6.37\, g\, \frac{Q(T_s - T_a)}{u^3 T_a}\, Z, \qquad (3-15)$$

where ΔH is in feet, u and v_s are in feet per second, Q in cubic feet per second at T_s, T_s and T_a are in degrees Kelvin, g = acceleration of gravity = 32.17 ft/sec^2, Z = ln J^2 + 2/J - 2, J = u^2//(Qv$_s$){0.43/(T$_a$/gG) - 0.28 v$_s$T$_a$/[g(T$_s$-T$_a$)]} +1, and G = potential temperature gradient ($^\circ$C/ft) = 0 for neutral, 0.003 for moderately stable (isothermal), and 0.006 for very stable (strong inversion). The terms are easily recognizable; the first is the velocity rise (ΔH_v) and the second is the temperature rise (ΔH_t). At distances downwind (x) where less than the maximum rise prevails, the rise may be calculated from

$$\Delta H_{vx} = \Delta H_v (1 - 0.8 \, \Delta H_v/x), \qquad x > 2\Delta H_v, \qquad (3-16)$$

and

$$x = 3.57 Y\sqrt{Qv_s/u} \ , \qquad\qquad\qquad (3-17)$$

where Y is used to calculate the Z value to use in ΔH_{tx} by Z = 4.17 log Y - 1.17 for Y > 5 and Z = 0.36 Y for Y < 5.

EXAMPLE 3-2: Compare the maximum plume rises calculated by the four methods presented for a 10-ft diameter stack emitting stack gas (essentially air) at 33 fps and 240°F into a 10-mph wind under slightly stable conditions and an ambient temperature of 60°F.

Holland formula (modified by Thomas):

v_s = 33 fps = 22.5 mph

Q_h = 33 π/4 (10)^2 34 gm/ft^3 (530°R/700°R) (0.24 cal/gm-$^\circ$
X (240°F - 60°F) 5/9 = 16 X 10^5 cal/sec

ΔH = [1.5(22.5)(10) + 6 X 10^{-4}(16 X 10^5)]
= (338 + 960)/10 = 130 ft

Carson and Moses formula:

A = 1 v_s= 10 m/sec   d = 3.05 m Q_h = 160   u = 4.47 m/sec

ΔH = 1/4.47 m/sec [4.12 (10)(3.05) + 9.22$\sqrt{160}$]

= 54.3 m = 178 ft

Bryant-Davidson formula:

v_s = 22.5 mph   u = 10 mph   d = 10 ft T_s = 116°C

T_a = 16°C

$\Delta H = 10 \, (22.5/10)^{1.4} \, [1 + (116 - 16)/116] = 58$ ft

CONCAWE formula:
$\Delta H = 0.175 \, (1.6 \times 10^6)^{1/2}/(4.47)^{3/4} = 72$ m $= 236$ ft

Bosanquet et al. formula:
$Q = 25,900$ cfs $u = 14.7$ fps $J = 10.3$ $Z = 2.85$

$$\Delta H = \frac{4.77}{1 + 0.43(14.7/33)} \sqrt{[25900(33)/14.7]}$$

$$+ \, 6.37 \, (32.2) \, (25,900) \, (100)(2.85)/[14.7^3 \, (289)]$$

$$= 305 + 1650 = 1955 \text{ ft}$$

at $x = 2000$ ft, $Z = 2.43$ and $\Delta H = 1112$ ft

at $x = 1000$ ft, $Z = 1.32$ and $\Delta H = 691$ ft

The disagreement of the various methods is quite evident in Example 3-2. If a smaller stack or a higher wind speed had been chosen, the Bosanquet et al. value would have been lowered faster than the others. Note the third power of the wind speed in the thermal rise term.

3. Sutton's Dispersion Formula

Sutton's formula has long been the most widely used model for plume dispersion. It is being displaced to a large extent by the more empirically based formulas of Cramer and Pasquill, Meade, and Gifford but is still frequently employed. Sutton's formula for a continuous point source is

$$X = \frac{2Q \, 10^6}{\pi C_y^2 C_z^2 x^{2-n}} \exp\left[\frac{-1}{x^{2-n}} \left(\frac{y^2}{C_y^2} + \frac{H_e^2}{C_z^2}\right)\right], \qquad (3\text{-}18)$$

where X = ground level concentration (ppm_v), Q = emission rate (ft^3/sec), x = distance downwind from stack (or virtual origin, see Fig. 3-10) (ft), y = distance crosswind from center line of plume (ft), H_e = effective stack height (ft), and all other terms are as defined for Eq. 3-6 (see Table 3-1 for constants). Any dimensional system works as long as it is consistent. For calculating other than ground concentrations, replace H_e with z, where z equals the vertical distance from the center line of the plume to the point of interest.

The location of the maximum ground concentration calculated by Sutton's equation is at $y = 0$ and

$$x_m = \left(H_e^2/C_z^2\right)^{1/(2-n)}, \qquad (3-19)$$

where the concentration is

$$X_m = 2 Q_m C_z/(e\pi u H_e^2 C_y). \qquad (3-20)$$

Some investigators approximate Eq. 3-19 to H_e/C_z. This is a poor practice as is evidenced by an example calculation.

EXAMPLE 3-3: What is the difference between the value from Eq. 3-19 and the approximation above when $n = 0.33$, $C_z = 0.1$, and $H_e = 100$ m?

$$(100^2/0.1^2)^{1/1.67} = (10^6)^{0.6} = 3980$$

$100/0.1 = 1000$ or about $1/4$ actual value.

Thus while the 0.6 power seems close to the 0.5 power, it is actually off by a factor of 4 which is intolerable in any reasonable estimate.

Numerous observers have disagreed with Sutton's equation. Some have tried to develop formulas of their own, while others have tried to modify the parameters of fit, the n and the C values. Holland suggested that gustiness (g_x) in Eq. 3-9 be replaced by the tangent of the standard deviation of the fluctuations in the azimuth angle (tan σ_a); the concept also holds for g_y in isotropic assumption. This concept is similar to that developed by Cramer (see Section V, B, 5). The meteorology group at Brookhaven National Laboratory has been very active in this field. Singer and Smith classified winds into types according to the traces of the azimuth on strip charts (see Fig. 3-11), then listed the Sutton coefficients by wind types (see Table 3-2). Many other investigators have made observations concerning the accuracy of the method; the literature is replete with such reports.

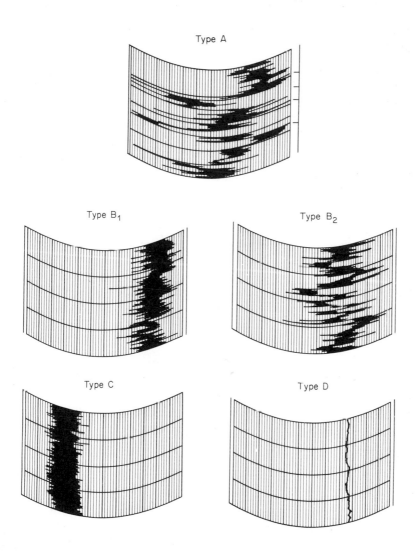

FIG. 3-11. Wind types by Smith and Singer.

TABLE 3-2

Experimental Values of Sutton's Diffusion Parameters[a,b]

Gustiness category	B_2	B_1	C	D
$C_y(m^{n/2})$	0.56	0.50	0.45	0.44
$C_z(m^{n/2})$	0.58	0.46	0.32	0.05
n	0.19	0.28	0.45	0.58
u_{8m}(m/sec)	2.5	3.4	4.7	1.9
u_{198m}(m/sec)	3.8	7.0	10.4	6.4
σ_y(m)	$0.40x^{0.91}$	$0.36x^{0.86}$	$0.32x^{0.78}$	$0.31x^{0.71}$
σ_z(m)	$0.41x^{0.91}$	$0.33x^{0.86}$	$0.22x^{0.78}$	$0.06x^{0.71}$

[a]By Singer and Smith, see Bibliography.

Type of surveys	Range, x, (m)	Sampling time (min)	Condition	C_y	C_z
Vertical	150-1000	15-20	Unstable	0.15 (0.12)	0.25 (0.19)[c]
			Neutral	0.09 (0.07)	0.20 (0.15)
			Stable	0.09 (0.07)	0.11 (0.08)
Ground	590-620	40	Neutral/ unstable	0.46 (0.35)	-- --
	880-1050	40	Neutral/ unstable	0.28 (0.22)	0.32 (0.25)
	1200	40	Neutral/ unstable	0.33 (0.25)	-- --
	2400-2800	15	Neutral/ unstable	0.23 (0.18)	-- --
	6000-9700	15	Neutral/ unstable	0.16 (0.12)	0.17 (0.13)
	1000-10000	60	Stable	<0.04	--

[b]By Stewart, Gale, and Crooks, see Bibliography

[c]Numbers in parentheses are equivalents for 3-min sample.

4. Pasquill-Meade-Gifford Method

A method proposed by Pasquill and by Meade and modified by Gifford is coming into dominance. The formula used is

$$X = \frac{Q}{\pi \sigma_y \sigma_z u} \exp\left[-\tfrac{1}{2}\left(\frac{y^2}{\sigma_y^2} + \frac{H_e^2}{\sigma_z^2}\right)\right], \tag{3-21}$$

where σ_i = standard deviation of the spread of plume concentration (σ_y horizontally and σ_z vertically). The statistical basis of the equation is again evident in the nature of the exponential term. The downwind distance (x) does not appear in the formula, but the diffusion parameters are functions of x. For plotted values of σ_y and σ_z, see Fig. 3-12.

5. Cramer's Method

Cramer used the same formulation as Gifford but employed spectral considerations to link σ_y and σ_z to the standard devia-

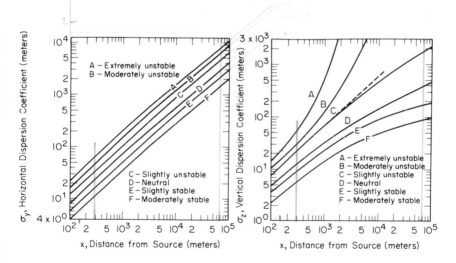

FIG. 3-12. Gifford dispersion coefficients. From F. A. Gifford, Jr., "Use of Routine Meteorological Observations for Estimating Atmospheric Dispersion," Nuclear Safety, 2:4, 48, June 1961.

tions of the wind vane fluctuations in azimuth and elevation (σ_a and σ_e), respectively, by

$$\sigma_y = \sigma_a x^f \quad \text{and} \quad \sigma_z = \sigma_e x^g, \tag{3-22}$$

where g ranges from 0.7 to 0.9 and f from 0.6 to >1.1 as the stability goes from extremely stable to extremely unstable (see Fig. 3-13 and Table 3-2). The σ_i values should be run over a time period of 20 min to satisfy the spectral considerations of this theory.

6. Bosanquet and Pearson Formula

The formula developed by Bosanquet and Pearson has been widely used and is rather easy to apply despite its formidable look. Their equation is

$$X = \frac{Q}{\sqrt{2\pi}\, pqux^2}\, \exp\left[-\left(\frac{y^2}{2q^2x^2} + \frac{H_e}{px}\right)\right], \tag{3-23}$$

where all terms are as before except that p = vertical diffusion coefficient and q = horizontal diffusion coefficient. For low, average, and moderate turbulence conditions, the values of p are 0.02, 0.05, and 0.10, respectively. Similarly, the q values are 0.04, 0.08, and 0.16.

The maximum ground concentration occurs at $y = 0$ and

$$x_m = H_e/2p, \tag{3-24}$$

with a value of

$$X_m = 0.215\, Q_m p / quH_e^2. \tag{3-25}$$

EXAMPLE 3-4: Calculate the downwind ground concentration on the centerline of a plume at 600 m from a stack with an effective height of 60 m discharging 100,000 scfm with 1200 ppm_v of sulfur dioxide under neutral stability conditions when the wind speed is 3 m/sec.

Sutton formula:
$$C_y = C_z = 0.11 \quad n = 0.25 \quad \text{(from Table 3-1)}$$
$$Q_v = 1200(100,000)(10^{-6}) = 120 \text{ cfm} = 2 \text{ ft}^3/\text{sec}$$

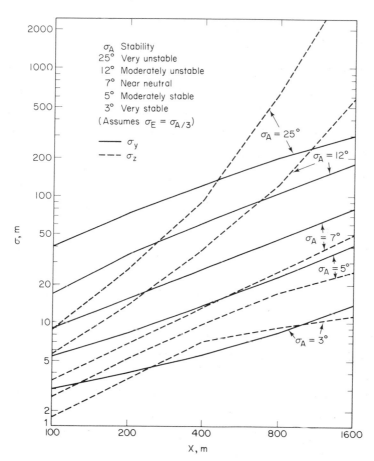

FIG. 3-13. Cramer dispersion coefficients. From H. E. Cramer, "A Brief Survey of the Meteorological Aspects of Atmospheric Pollution," <u>Bull. Am. Meteorol. Soc</u>., <u>40</u>:4, 169, April 1959.

$$X = \frac{2\ (2)\ 10^6\ (1/35.3)}{\pi\ (0.11)^2\ 3\ (600)^{2-0.25}}\ \exp\left[-\frac{1}{600^{1.75}}\left(\frac{60^2}{0.11^2}\right)\right]$$

$$= 13.6\ (0.0169) = 0.23\ \text{ppm}_v$$

Gifford method:

$$\sigma_y = 45\text{m} \qquad \sigma_z = 22\ \text{m} \qquad \text{(from Fig. 3-12)}$$

$$Q_m = 2/35.3\ (64,000/24) = 152\ \text{gm/sec}$$

$$X = 152/[\pi(45)(22)(3)] \exp[-1/2(60^2/22^2)]$$
$$= 0.0163(0.024) = 0.00039 \text{ gm/m}^3 = 0.15 \text{ ppm}_v$$

Cramer method:

$$\sigma_y = 36 \text{ m} \qquad \sigma_z = 19 \text{ m} \qquad \text{(from Fig. 3-13)}$$
$$X = 2(152)/[\pi(36)(19)(3)] \exp[-1/2(60^2/19^2)]$$
$$= 0.0473(0.0068) = 0.00032 \text{ gm/m}^3 = 0.12 \text{ ppm}_v$$

Bosanquet and Pearson formula:

$$p = 0.05 \qquad q = 0.08 \qquad \text{(average turbulence)}$$
$$X = 152/[\sqrt{2\pi}(0.05)(0.08)(3)(600^2)]\exp[-60/0.05(600)]$$
$$= 0.0141(0.135) = 0.0019 \text{ gm/m}^3 = 0.714 \text{ ppm}_v$$

It should be noted that the downwind distance is x_m for the Bosanquet and Pearson formula; the comparable point (1350 m) for Sutton's shows 0.0033 gm/m^3 or about 1.2 ppm$_v$.

C. Line Sources

Line source geometry is normally used for a line of discrete continuous point sources, and then the concept is valid only if the wind is blowing across rather than along the source line. Several industrial and power plants have stacks that are lined up and fit this category; it is also applicable to a long-swing or stationary irrigation pipe applying an entrained pesticide or fertilizer or to a line of simultaneous explosions (nuclear or nonnuclear) used to excavate a canal, provided the instantaneous point source formula is used as a starting basis.

A continuous line source differs from a continuous point source because lateral dispersion replaces the concentrated material from one plume by like material from an adjacent plume; therefore the only effective dispersion (until some number of line lengths downwind) is vertical dispersion. Sutton's equation is modified by omitting the y and C_y terms and using the square roots of π and x^{2-n} terms in the denominator of the part of the equation in front of the exponential. Similar reasoning can be applied to the other dispersion formulas.

D. Areal Sources

Large industrial complexes and urban areas may be considered as a single source of air pollution for some analytical purposes. This is especially appropriate if their emissions are alike or similar enough that each individual source contributes to the overall problem. Models have generally taken the form of considering the box that would contain the source with a height equal to the mixing depth and a cross section that is the rectangular equivalent of the source area. The ventilation rate into the box and the equilibrium times with complete mixing are reviewed more fully by Wanta. A British research report for Leicester says that smoke from an areal source diminishes as the distance for the first 4-10 mi, as the distance squared from 10 to 100 mi, then slowly again at distances greater than 100 mi. One logical approach seems to be the assumption of an exponential decay, then estimate a decay constant or half-life (see Fig. 3-14) based on the removal of the pollutant by chemical change, fallout, washout and rainout, and absorption or impaction on surfaces.

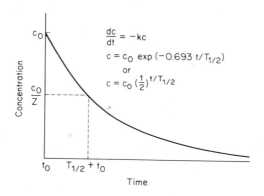

FIG. 3-14. Exponential decay curve.

The type of analysis most used is probably the superposition of the contributions from individual sources, or putting all the sources in a grid area (1 mi^2) at a point so that the many sources are classed into several for analytical purposes. The contribution from each source can then be determined by use of a suitable model and the individual contributions summed to obtain the estimated concentration at any point. Such techniques are prohibitive by any means other than computer analysis.

E. Limitations of Dispersion Calculations

It must be kept in mind that dispersion models assume that the concentrations across plumes, both horizontally and vertically, follow a Gaussian distribution. While average concentrations may follow this distribution closely, instantaneous concentrations fluctuate widely because of plume movements in the vertical (see Fig. 3-9) and in the horizontal (see Fig. 3-15) and nonisotropic mixing. Occasionally, laminar flow persists and a plume may travel for miles essentially unchanged in size or concentration (see Fig. 3-16).

Sutton's and Cramer's methods are usually considered to apply only over short distances, up to about 1 km. Gifford recommends his method for distances up to 50 km or more. Presumably, the right dispersion coefficients for any given distance and formula permit the use of the formula at that distance.

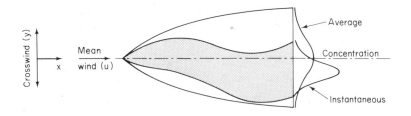

FIG. 3-15. Plume streakline and average path.

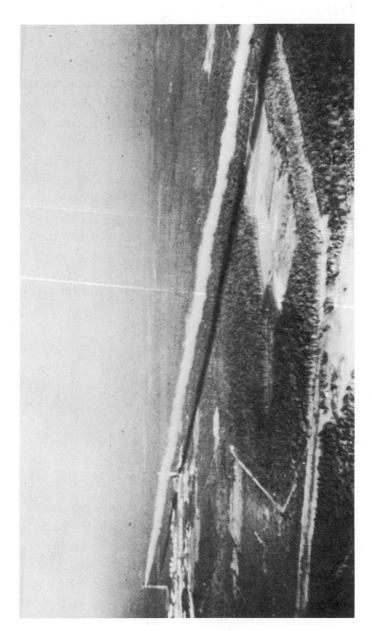

FIG. 3-16. Nonmixing plume during inversion. From
M. E. Smith, "Reduction of Ambient Air Concentrations of
Pollutants by Dispersion from High Stacks," Proc. 3rd Natl.
Conf. Air Pollution, Washington, D. C., December 12-14, 1966.

The models differ considerably in calculated magnitudes.
One explanation used for these variations is to say that Sutton's
gives a 3-min average peak concentration and Bosanquet and
Pearson's a 30-min average (see Chapter 6, Section II, C);
however, it is still necessary to explain the wide discrepancies
of the distance from the source to the maximum ground concentration. The real trouble is that despite several experiments
conducted in attempts to describe atmospheric transport in terms
of a suitable model and parameters, many more will have to be
made to formulate the actions of such a widely varying medium.
In the meantime the concentrations calculated should be viewed
as orders of magnitude and thus serve a very valuable purpose
until better methods are available.

VI. PARTICLE DEPOSITION

Particles are removed from the atmosphere by sedimentation,
washout and rainout, and impaction on surfaces of trees, structures, and other objects in contact with the air. The time of
removal varies with particle size, height above the surface, and
the meteorological factors that transport the particles. Some
particles are airborne only a very short time (measured in seconds)
while others remain airborne for long periods (up to years), such
as the very small fallout ash (containing ^{90}Sr) injected into the
stratosphere by nuclear explosions or the small dusts from the
eruption of Krakatoa.

A. Sedimentation

The dispersion models may be used for downwind concentrations of particles by assuming an inclined plume and replacing
H_e with $H_e - x \tan (u_t/u)$, where u_t is the terminal settling velocity of the particles as described in Chapter 5, Section II, G, 4.
For small angles the tangent of an angle is approximately equal
to the angle and $x \tan (u_t/u)$ may be used as xu_t/u. The rate of
deposition is calculated by multiplying the concentration by the
settling velocity.

Chamberlain modified Sutton's formula to fit dry deposition by including a velocity of deposition (v_g) such that

$$X = \frac{Q}{\pi C_y C_z u x^{2-n}} \exp\left[-\left(\frac{4v_g x^{n/2}}{nuC_z\sqrt{\pi}} + \frac{y^2}{C_y^2 x^{2-n}} + \frac{H_e^2}{C_z^2 x^{2-n}}\right)\right]$$

(3-26)

There should be no reflection of particles; therefore the 2 in the numerator is omitted. The v_g parameter has been measured for ^{131}I fallout as 0.1-2.8 cm/sec for various conditions. The measurements confirmed the diffusion nature of removal because the amount of material on the bottom of the trays was 70% of that on the top.

B. Washout and Rainout

Particles may be removed by precipitation from either impaction on water droplets or condensation around the particle to form a droplet. The first mechanism is called washout; the second, rainout. Some analyses have shown up to 25 particles per drop of rain for normal dust concentrations. Under unusual conditions, the rainfall in dust storm country consists merely of drops of mud, and there have been black rains of soot and red snows with iron oxide particles around steel mills.

Chamberlain defined a parameter (Λ) of deposition rate for washout--the proportion of cloud removed per second (see Fig. 3-17). This factor can be included in Sutton's equation by replacing the first term of the exponential in Eq. 3-26 with $\Lambda x/u$. The rate of deposition may be obtained from this modified equation by setting H_e equal zero and multiplying the concentration obtained by Λ.

C. Impaction

Impaction is quite important in the removal of small particles from the atmosphere, especially those particles small enough to prevent effective settling. These particles are also

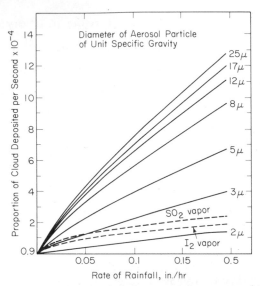

FIG. 3-17. Values of deposition parameter (Λ). From
A. C. Chamberlain, <u>Aspects of Travel and Deposition of Aerosol
and Vapour Clouds</u>, A.E.R.E. Rept. HP/R 1261, Harwell, Great
Britain, 1953.

resistant to washout. The velocity needed for impaction is given
to the particles by air currents, and for very small particles
(<0.2 µm) by Brownian motion. The efficiency of removal depends
upon the diameter of the particle, the characteristic length (diameter) of the impaction surface, and the relative velocity between
the particle and the surface (see Chapter 5, Section II, G, 6) for
more extensive discussion of this subject). The most effective
impaction occurs when an aerosol is blown through tree leaves
and grass or weeds. Impactive removal of particles from the air
causes the vertical and even the overhead surfaces of structures
to become dirty.

Particles also collide with each other to form larger particles
which can then settle, but this is a slow process below a particle
concentration of about 10^7 particles/cm^3, a concentration far
above that usually found in the atmosphere but which may be
reached in dusty situations such as plumes.

VII. GASEOUS REMOVAL

Gases are removed from the atmosphere by chemical reaction with other gases, through absorption in precipitation or water on the surface of objects and the earth, and through adsorption on airborne and surface solids. The dispersion models for the nonconservative gases can be modified by putting in a new term which accounts for exponential (first-order) decay, a term such as exp (-kt) or exp $(0.693t/T_{\frac{1}{2}})$. The decay constant (k) or the half-life $(T_{\frac{1}{2}})$ requires that considerable care and judgment be exercised in its selection. The t may be of the form x/u. A more extensive discussion of the removal of gaseous pollutants is included in Volume II of this treatise.

A. Chemical Reactions

Most of the important atmospheric reactions of pollutants involve oxidation. The large concentration of oxygen tends to drive these reactions at significant rates, whereas the very dilute gases usually react very slowly, and nitrogen is rather inert. The principal point in the chemical removal of a pollutant gas is its reaction rate. Unfortunately, many of these rates are not accurately known. Some of the reactions most relevant to the analysis of atmospheric transport of air pollution are the oxidation of sulfur dioxide and the oxides of nitrogen.

The lower oxidation forms of nitrogen go to nitrogen dioxide (NO_2), which oxidizes to nitrogen pentoxide that rapidly reacts with water vapor to form nitric acid. The reaction rate constant for the removal of nitrogen dioxide is probably on the order of 4×10^7 cm^3/mole-sec.

The lower oxidation forms of sulfur are oxidized to sulfur trioxide, which reacts with water vapor to form sulfuric acid. A reaction rate of the order of 0.1%/hr in strong sunlight has been reported. These reactions and those for the oxides of nitrogen are driven photochemically, by photons ordinarily found in sunlight.

Not all of the atmospheric chemical reactions result in the reduction of pollution. The Los Angeles smog problems are compounded by substances, such as PAN, which are products of photochemical reactions. The color of the smog is caused by nitrogen dioxide.

B. Absorption

Soluble gases are quite readily removed by absorption into rain or other precipitation, water on the surface of plants and structures, or bodies of water. Sometimes they are removed by absorption in mists other than water. The controlling factors in absorptive removal are the solubility of the gaseous pollutant, the driving pressure, the rate of renewal for the surface, and the area of contact between the gaseous and the liquid phases. Gas transfer coefficients are available in absorption tower design data and can be used for estimating the decay constant or half-life of an atmospheric reaction. Because the process data are for concentrations higher than those found in ambient air, Henry's constant may be of more benefit than any other data.

The same decay term used for the rainout of particulates can be used for precipitation removal of gas. The values of the decay constant (Λ) are shown in Fig. 3-17.

C. Adsorption

Adsorptive removal of gases is usually small compared with chemical reaction and absorptive removals; however, very dusty atmospheres may effect significant cleansing of a gaseous pollutant by the removal of the particles with their adhering gases. Any adsorption calculations made on atmospheric pollutants would be gross estimates, probably made with an adsorption isotherm equation such as the Freundlich (Küster) isotherm (Chapter 5, Section I, G).

VIII. APPLICATIONS

The principles and techniques of atmospheric transport of pollutants are highly important to air pollution analysis. The degree of importance far outweighs the accuracy of the methods because an answer with a factor of approximately 2 is usually sufficient for the knowledge of effects and factors of safety which are applied. Transport calculations are needed for land-use planning, especially for plant siting of any major pollution source, and are required for locating nuclear power plants.

A. Downwind Concentrations and Dose

Knowledge of downwind concentrations and the dose at a point can aid in the prevention of air pollution problems, but it is also important in the analysis of air pollution. Dose means concentration multiplied by time of exposure. It is usually much easier to employ stack sampling and calculate the expected downwind concentrations that are likely rather than to measure the concentrations; this procedure is practical because of the number of samples required to describe the downwind plume and the limited accessibility of areas from which the samples are desired. The calculated concentrations may be plotted in the manner of a wind rose to show the concentration when the wind is from a given direction (see Fig. 3-18).

B. Isodose Maps

The dispersion models used in conjunction with wind data can yield the dosage of pollutants at given distances and directions from the source. The points of determined dose are then used to draw isodose contours (lines of equal dose) for a pollution map of the area (see Fig. 3-18b). The units of dose may be parts per million hours, micrograms per cubic meter hours, milliroentgens, picocuries per cubic meter hours, and so on. Often the wind data available are not adequate for a complete analysis,

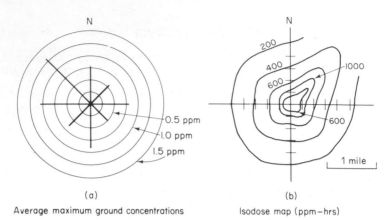

(a)   (b)

Average maximum ground concentrations   Isodose map (ppm—hrs)

FIG. 3-18. Pollutant dispersion presentations. (a) Average
maximum ground concentrations. (b) Isodose map (parts per
million-hours).

but a simple wind rose (see Fig. 3-5) can be used by stipulating
the stability conditions likely for each wind velocity range.

An isodose map is quite useful in analyses made for plant
location, accidental releases of pollutants (particularly releases
of radionuclides), calculating the effects of overlapping sources,
and analysis of an existing problem source.

Isodose maps were used to predict fallout from the nuclear
detonations set off in Nevada and the Pacific. Twelve Rawinsonde
stations surrounded the Nevada test site during some of its operations. The trajectories of the cloud and its fragments were predicted for each blast, the amount of expected dispersion calculated, the rate of deposition estimated, and the amount of predicted fallout plotted on a map of the area.

C. Forecasting Air Pollution Potential

Meteorologists in the weather research group of the national
air pollution control effort have been making forecasts of air pollution potential since August 1960. The ventilation rate or total
dilution expected in areas of stagnation forms the basis of these
forecasts. The factors involved are mixing height, wind speed,

and shear. These factors depend upon the location, movements, and gradients of high-pressure fields and the cloud cover or the insolation. Forecasts for air pollution potential serve the analytical phase of air pollution in planning samplings of adverse conditions, but the principal utility of such forecasts is in prevention, especially for planning operations that depend upon dilution of the emissions to prevent serious problems. Air pollution potential forecasts are routinely delivered over the weather teletype service.

IX. TRANSPORT OTHER THAN WIND-CARRY

Air carries noise and permits the passage of electromagnetic radiations (photons) and, to a limited extent, projected particles, even in the absence of winds and molecular diffusion. All of these pollutants are absorbed by the air within some finite distance, which is analogous to the decay of wind-carried pollutants.

A. Noise Carriage

The presence of air is essential to the production of sound as indicated by the definition of sound--the effect that fluctuating air pressures have on the ear. The velocity of propagation for sound in air is about 331.5 m/sec at STP.

Because of the velocity variation with temperature, sound seldom travels in straight lines. In an atmospheric lapse the refraction of sound is upward; in an inversion it is downward. In addition to the refraction effect, noise may be reflected from layers where temperature and winds change markedly. The attenuation of noise varies approximately with the square of the distance.

B. Electromagnetic Radiation Transmission

Electromagnetic radiation penetrates the air until a collision occurs between the photon and the electron field of an atom. The average distance traveled before such a collision is the mean-free-path or relaxation length for the radiation. The interactions are normal and may be represented by the first order reaction.

Not all interactions result in the removal of the photon; some interactions of high energy photons give an attenuated photon and a displaced or excited electron. If the total energy remaining is desired rather than the number of unchanged incident photons, a buildup factor must be used to account for the scattered photons.

$$E = E_0 B e^{-\mu s}, \tag{3-27}$$

where E = photon energy at distance s (number of photons X energy per photon), E_0 = initial photon energy, μ = total attenuation coefficient (about 0.00004 cm^{-1} for 6-MeV gammas and 0.0002 cm^{-1} for 0.2 MeV gammas or x rays), and B = buildup factor (about $1 + \mu s$ for air).

The scatter and absorption of less energetic photons are discussed in Chapter 10, Section III, A, 4.

C. Energetic Particle Transmission

Particles projected in the air with considerable energy lose their energy within a short distance--a micrometer-sized particle within millimeters and a β particle within meters. The stopping of particles of micrometer size follows the dynamic laws of drag and inertia. Charged particles such as β particles react with the electron fields of atoms in the air and are slowed.

X. SUMMARY

The atmosphere is in constant movement. Air movements carry air pollutants, gases, and small particles, along with its own gases. The infrastructure of the air currents or winds can only be described statistically because of the fluctuations in both speed and direction. The normal wind structure has turbulent eddies or cells superimposed on the average velocity vector. Eddies are described by intensity and scale as determined from recording traces of wind speed and direction. Eddies cause transport in directions normal (horizontal and vertical) to the average wind.

Most air pollution comes from continuous point sources, resulting in plumes. The concentrations across a plume are usually assumed to be Gaussian; therefore the dispersion models have a term of exp $[-(\text{distance})^2/(\text{length parameter})^2]$. Simple logic places the emission amount in the numerator and the wind speed in the denominator of a dispersion formula. These factors form the basis of all statistical dispersion models described in this chapter. The order of magnitude values for the expected dispersion with downwind distance are shown in Fig. 3-19.

Although dispersion formulas are set up to calculate down-wind concentrations (particularly those at the ground) for gases, the formulas may be used for particles by considering the plume to be inclined. Account should be taken of the decrease in con-centrations caused by processes of removal as well as dispersion. These processes include dry deposition from sedimentation of particles or adsorption of gases and removal of both particles and

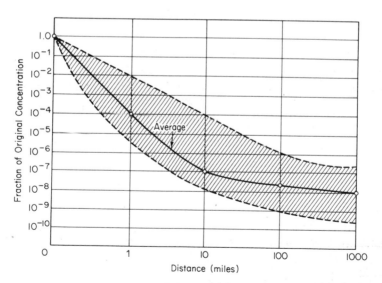

FIG. 3-19. Atmospheric dispersion. From Waste Manage-ment and Control, National Academy of Sciences--National Research Council, Publ. 1400, Washington, D. C.

gases by washout, rainout, and absorption. An extra term in the transport equation may correct for decreases caused by the non-conservative nature of the pollutants.

PROBLEMS

1. What is the probable daytime wind speed at a height of 2 m when a 10-m wind is 12 knots?

2. Draw a lapse diagram and indicate the stability for each section if the lapse data are: 0 ft, 53°F; 100 ft, 62°F; 400 ft, 59°F; 1000 ft, 56°F; 1500 ft, 56°F; 2500 ft, 54°F; and 6000 ft, 35°F.

3. If the lapse in problem 3 prevails at 2 a.m. and the forecast maximum temperature for the day is 77°F, what will be the mmd forecast?

4. Assuming uniform concentration over the whole cross section and no deposition or decay and using a 10-mph wind, an mmd of 1200 ft, and a shear angle of 5°, calculate the dispersion expected for a 400,000-cfm plume in a distance of 10 mi. Compare with the Fig. 3-19 value.

5. Calculate the concentration expected in problem 4 by Gifford's method, assuming a moderately stable atmosphere and an effective stack height of 60 m.

6. Distribute the calms of the wind rose (Fig. 3-5) to each direction in proportion to the fraction of the low-speed winds (<8 mph) in that direction. Assume that winds ≤3 mph are during stable conditions; 4-7 mph, moderately stable; 8-12 mph, neutral; 13-18 mph, unstable; and >18 mph, highly unstable. With these conditions determine: (a) the downwind concentration curve for each wind speed from the southeast; (b) the downwind dose curve in each case; (c) the sum of the dose curves to give the overall curve from the source to the northwest. Use a 100-m effective stack height, sulfur dioxide emission of 300 tons/day, and Sutton's equation.

BIBLIOGRAPHY

Air Pollution Manual, Part I: Evaluation, American Industrial Hygiene Association, Detroit, Michigan, 1960.

R. D. Cadle and P. L. Magill, "Chemistry of Contaminated Atmospheres," Air Pollution Handbook (P. Magill, F. Holden, and C. Ackley, Eds.), McGraw-Hill, New York, 1956.

S. B. Carpenter et al., "Full-Scale Study of Plume Rise at Large Coal-Fired Electric Generating Stations," J. Air Pollution Control Assoc., 18:7, 458-465, July 1968.

J. E. Carson and H. Moses, "The Validity of Several Plume Rise Formulas," J. Air Pollution Control Assoc., 19:11, 862-866, November 1969.

A. C. Chamberlain, Aspects of Travel and Deposition of Aerosol and Vapour Clouds, A.E.R.E. Rept. HP/R 1261, Atomic Energy Research, Harwell, Great Britain, 1953.

P. E. Church, "Dilution of Waste Stack Gases in the Atmosphere," Ind. Engr. Chem., 41, 2753-2756, 1949.

F. A. Gifford, Jr., The Problem of Forecasting Dispersion in the Lower Atmosphere, U.S.A.E.C. Report of Collection of Articles Published in Nuclear Safety in 1960-1961, Oak Ridge, Tennessee.

C. A. Gosline, L. L. Falk, and E. N. Helmers, "Evaluation of Weather Effects," Air Pollution Handbook (P. Magill, F. Holden, and C. Ackley, Eds.), McGraw-Hill, New York, 1956, Section 5, 5-1 through 5-66.

C. R. Hosler, "Low-Level Inversion Frequency in the Contiguous United States," Monthly Weather Rev., 89, 319-339, September 1961.

G. C. Holzworth, "A Study of Air Pollution Potential for the Western United States," J. Appl. Meteorol., 1:3, 366-382, September 1962.

P. A. Leighton, Photochemistry of Air Pollution, Academic Press, New York, 1961.

R. A. McCormick, "Air Pollution Climatology," Air Pollution, 2nd ed. (A. C. Stern, Ed.), Academic Press, New York, 1968, Vol. I, Chapter 7, 187-226.

Meteorological Aspects of Air Pollution, Training Course Manual, U.S.P.H.S., D.H.E.W., Cincinnati, Ohio, 1964.

Meteorology and Atomic Energy, U. S. Weather Bureau (for U.S. A.E.C.), Washington, D. C., July 1955.

F. Pasquill, Atmospheric Diffusion, D. Van Nostrand, London, 1962.

I. A. Singer and M. E. Smith, "Relation of Gustiness to Other Meteorological Parameters," J. Meteorol., 10:2 121-126, 1953.

D. H. Slade, Ed., Meteorology and Atomic Energy, 1968, E.S.S.A. (for U.S.A.E.C.), Washington, D. C., July 1968.

M. E. Smith, "The Forecasting of Micrometeorological Variables,"
    Meteorol. Monographs, 1:4, 50-55, 1951.

N. G. Stewart, H. J. Gale, and R. N. Crooks, "The Atmospheric
    Diffusion of Gases Discharged from the Chimney of the
    Harwell Reactor BEPO," Intern. J. Air Pollution, 1, 31-43,
    1958.

O. G. Sutton, Micrometeorology, McGraw-Hill, New York, 1953.

F. W. Thomas, S. B. Carpenter, and W. C. Colbaugh, "Plume
    Rise Estimates for Electric Generating Stations," J. Air
    Pollution Control Assoc., 20:3, 170-177, March 1970.

R. C. Wanta, "Meteorology and Air Pollution," Air Pollution,
    2nd ed. (A. C. Stern, Ed.), Academic Press, New York,
    1968, Vol. I, Chapter 7, 187-226.

Chapter 4

STATISTICS FOR AIR SAMPLING AND ANALYSIS

I. INTRODUCTION

The only sizes that are exact are those that are defined; for example ^{12}C has an atomic weight of 12 with as many zeros as desired following the decimal. All measured sizes are only approximate. Errors of measurement are termed systematic errors if they are known in size and direction and accidental errors if they are not known. Because it is highly desirable to be able to give some information about the probable size of accidental errors, a quite useful mathematical representation--error statistics--has been developed.

II. FREQUENCY HISTOGRAMS AND FREQUENCY CURVES

When many measurements of a variate are made, grouped according to size, and plotted as a bar graph, a representation such as Fig. 4-1a is obtained; this is a frequency histogram. The measurements lose exactness from the grouping by size, but this procedure is warranted in the interest of simplifying presentation of the data. It should always be kept in mind that the maximum amount of information known about data is contained in the original items without any statistical treatment.

If the number of measurements were increased without limit and the width of the bars in the histogram became infinitesimal, a frequency curve would result (see Fig. 4-1b). A frequency curve represents one further step in transformation of the data;

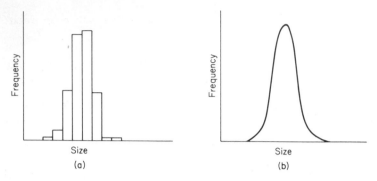

FIG. 4-1. Frequency plots. (a) Frequency histogram.
(b) Frequency curve.

however, a limited number of measurements would necessarily
fluctuate from the actual population values and the curve is a
good smoothing device.

III. FREQUENCY CHARACTERISTIC PARAMETERS

Certain parameters have been defined in an effort to characterize frequency arrays. Two types of parameters predominate--
those for central tendency and those for dispersion. If a distribution type is known (or assumed), a centering value and a measure of dispersion will rather fully define most distributions. The
measure of dispersion may require more than one parameter.

A. Centering Values

One logical centering value is the high point of the frequency plot (that size occurring most frequently); this value is
termed the mode ($\tilde{x}$) or modal size. The mode is the expected
value of a member taken from that population. Not all frequency
curves display a mode; some have no mode and are called J-shaped,
while others have two or more humps.

The size that divides the area under a frequency curve in
half is often used for the centering value--this is the median.
For discrete data the median is the central value in array ordered

by size, or between the two centralmost values when an even
number of items is in the array.

The most used, and quite often misused, centering value
is the mean or average. The mean or arithmetic mean is the
centroid of the frequency plot; a cutout of the frequency plot
would balance on a knife edge along the mean size. For continuous
data the mean (arithmetic) is calculated as

$$\bar{x} = \frac{\int_x x\, f(x)\, dx}{\int_x f(x)\, dx} \, , \tag{4-1}$$

where $\bar{x}$ = mean (arithmetic) of x, f(x) = frequency (probability
density) for given x, and integration is over the entire range of x.
For discrete data

$$\bar{x} = \sum_i x_i / N, \qquad i = 1 \text{ to } N \tag{4-2}$$

where $\bar{x}$ = mean (arithmetic) of x and N = number of x values in
the data array.

Other means often used are the geometric mean, the harmonic mean, the weighted mean, and the moving average. The
geometric mean is the Nth root of the product of all N values of
x. Another way of defining the geometric mean is that it is the
antilogarithm of the arithmetic mean of the logarithms of the
variates x. The harmonic mean is the reciprocal of the average
of the reciprocals of the values of the variate. To obtain the
weighted mean of a set of values, each value is multiplied by
its number of occurrences (or abundance) and the products summed;
the sum is then divided by the total occurrence (or abundance) in
the set. The weighted mean in the case of classified or grouped
data is the arithmetic mean. The moving average is used to
smooth data by averaging a fixed number of adjacent points with
each point in turn and plotting the average. The factors may be
weighted by position with the central value having the most
weight and the remote values the least weight for a particular
value.

B. Dispersion Values

The characterizing parameters usually employed with statistical distributions are functions of the moments of the set of values about their mean. Statisticians use the standardized moments (α) which are simply the moments of the transform of the variate x. The transformation is done by subtracting the x value from $\bar{x}$ and dividing by the standard deviation which is defined below. The regular moments about the mean without transformation will serve the purpose here. For continuous data the moments about the mean are defined by

$$\mu_k = \frac{\int_x x^k\, f(x)\, dx}{\int_x f(x)\, dx}\,, \qquad k = 0,\ 1,\ 2\ldots \qquad (4\text{-}3)$$

where μ_k = the kth moment of x about the mean and the other items are as before. For discrete data the moments about the mean are defined by

$$\mu_k = \sum_i (\bar{x} - x_i)^k /N\,, \qquad i = 1 \text{ to } N \qquad (4\text{-}4)$$

The zeroth moment (k = 0) equals 1 and the first moment (k = 1) equals 0. The second moment is used as a measure of dispersion; μ_2 is the variance (σ^2). The positive square root of the variance, that is, the root-mean-square deviation, is called the standard deviation (σ). Its geometrical significance is that it is the moment of inertia of the area under the frequency curve. The standard deviation is actually calculated from

$$\sigma = \left[\ \sum_i (\bar{x} - x_i)^2 /(N - 1)\ \right]^{1/2}, \qquad i = 1 \text{ to } N \qquad (4\text{-}5)$$

where σ = the standard deviation of the variate x. (Note: 1 is subtracted from N to remove the degree of freedom taken from the data in defining its mean. This concept is rigorously treated in books on statistics.)

Sometimes merely the sum of the squares of the definitions is used as a measure of dispersion. Another occasionally used parameter of dispersion is the sum of the absolute values of the deviations from the mean. However, dispersion parameters are largely limited to the standard deviation or the variance.

C. Skewness

Some parameter is needed to denote the tendency toward symmetry about the mean, the amount that "the hat is cocked." Figure 4-2 shows zero skewness and positive skewness in parts a and b, respectively. Distributions that are symmetrical about the mean have zero skewness. Frequently used measures of skewness are calculated by

$$Sk_p = \mu_3/\mu_2^{3/2} , \tag{4-6}$$

and

$$Sk = (\bar{x} - \tilde{x})/\sigma . \tag{4-7}$$

The amount of skewness is difficult to interpret from the definitions above, particularly the Eq. 4-6. Sometimes Bowley skewness is used; it has the advantages of less dependence on the kurtosis (see next paragraph) and a range of 0 to 1.

$$Sk_b = \frac{\left[x_{75} - x_{50}\right] - \left[x_{50} - x_{25}\right]}{\left[x_{75} - x_{50}\right] + \left[x_{50} - x_{25}\right]} , \tag{4-8}$$

where Sk_b = Bowley skewness and x_{ij} = ijth percentile size with data in increasing order. (x_{75} means that 75% of the variates are less than or equal to this size.)

D. Kurtosis

A less meaningful statistical parameter is the length and height of the tails of the frequency curve--the kurtosis. The fourth moment about the mean is taken as a measure of kurtosis. μ_4/μ_2^2 should have a value of 3 for the normal distribution. A

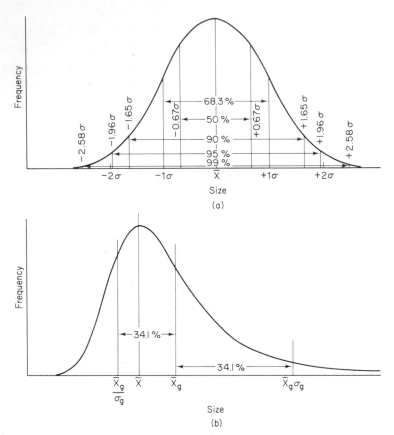

FIG. 4-2. Common frequency curves. (a) Normal distribution curve. (b) Log-normal distribution curve.

higher value means a flatter top on the frequency curve, and a lower value means more peakedness in the curve. One-half the difference between the 75 and 25 percentile values divided by the difference between the 90 and 10 percentile values is also used to measure kurtosis. This parameter has a value of 0.263 for the normal distribution.

EXAMPLE 4-1: The data shown in the tabulation represent the results of 300 samplings for suspended particulates. Calculate all of the previously defined statistical parameters for these values.

EXAMPLE 4-1

| Measured data | | | Derived values | | | | |
Group bounds ($\mu g/m^3$)	No. f	Midpoint m	fm	$m - \bar{x}$ d	fd^2	fd^3	fd^4
55– 64	1	59.5	60	-40.4	1,632	- 65,939	2,663,946
65– 74	7	69.5	456	-30.4	6,469	-196,661	5,978,502
75– 84	26	79.5	2,067	-20.4	10,820	-220,731	4,502,918
85– 94	60	89.5	5,370	-10.4	6,490	- 67,492	701,915
95–104	101	99.5	10,050	- 0.4	16	-	3
105–114	69	109.5	7,555	9.6	6,359	61,048	586,058
115–124	27	119.5	3,226	19.6	10,372	203,297	3,984,630
125–134	7	129.5	906	29.6	6,133	131,540	5,373,594
135–144	2	139.5	279	39.6	3,136	124,198	4,918,252
Totals	300		29,970		51,427	19,254	28,709,818

$x_{med} = x_{50} \approx 100.1 \ \mu g/m^3$ $x_{25} \approx 92 \ \mu g/m^3$ $x_{75} \approx 109 \ \mu g/m^3$ (estimated by interpolation)

$\sigma = [\Sigma fd^2/(\Sigma f - 1)]^{1/2} = \sqrt{(51,427/299)} = 13.1 \ \mu g/m^3$ Kurtosis $= 95,699/(172)^2 = 3.24$

$\mu_k = \Sigma fd^k/\Sigma f$ $\mu_0 = 1$ $\mu_1 = 0$ $\mu_2 = 172$ $\mu_3 = 64.1$ $\mu_4 = 95,699$

$Sk_p = 64.1/(172)3/2 = 0.028$ $Sk = (99.9 - 100.1)/13.1 = -0.015$

$Sk_b = [(109 - 101) - (101 - 92)]/(109 - 92) = -0.06$

IV. BINOMIAL DISTRIBUTION

The probability relations of any group of happenings that are always amenable to binary classification--success:failure, yes:no, go:no go, heads:tails, and so on--can be represented by the binomial distribution function.

$$P(x,n) = \frac{n!}{(n-x)!\,x!}\ p^x\,q^{n-x}, \qquad\qquad (4-9)$$

where $P(x,n)$ = the probability of x successes in n trials (x and n are non-negative integers), p = the probability of success of any trial, and q = the probability of failure = 1 - p by definition. The mean of the binomial distribution is np and the standard deviation is $\sqrt{npq}$.

EXAMPLE 4-2: What is the probability of rolling exactly three sixes out of five rolls of a single die? (A die is a cube with faces number 1 through 6.)

$$p = 1/6 \qquad q = 5/6 \qquad n = 5 \qquad x = 3$$

$$P(3,5) = \frac{5!}{(5-3)!\,3!}\ (1/6)^3\ (5/6)^{5-3} = 0.03215$$

The binomial distribution has its place in sampling statistics, as is shown later by example, and can be used to derive the Poisson and normal distributions.

V. POISSON DISTRIBUTION

When binomial distribution is applied to a population with a very small probability p (or q, since the roles can be reversed) and the number of trials n is large, the binomial probability can be approximated by the Poisson relationship:

$$P(x,n) = \frac{\lambda^x}{x!}\ e^{-\lambda}, \qquad\qquad (4-10)$$

where $P(x,n)$ = the same as before, λ = np (usually 0.1 to 10), and e = the base of natural logarithms = 2.71828+. The Poisson distribution has the peculiarity that the mean and the variance are equal.

EXAMPLE 4-3: What is the probability of throwing 12 on a pair of dice exactly two times in 100 throws?

$$p = (1/6)^2 \qquad n = 100 \qquad x = 2 \qquad \lambda = np = 25/9$$

$$P(2,100) = \frac{(25/9)^2 \, e^{-25/9}}{2!} = \frac{625 \times 0.062}{81 \times 2} = 0.241$$

Had it been desired to obtain the probability of rolling 12 at least twice in a 100 throws, the probability would have been the sum of the probabilities of rolling 12 2 times, 3 times... 100 times or 1 minus the probabilities of 0 times and 1 time $(1 - 0.062 - 0.241 = 0.697)$. Similar logic could be applied to Example 4-2.

VI. NORMAL DISTRIBUTION

Errors of measurement are random in sign, plus and minus being equally likely except for those measurements limited in one direction (near zero and so on); however, the size of the error is not random but has a greater probability of being small than large because of the skill involved in the measurement. The frequency curve for random errors is variously termed bell-shaped, Gaussian, or normal. Certain other populations exhibit similar frequency distributions; these include the lengths of time between successive disintegrations in a radioactive material, the heights of a given group of people, the numbers of red blood cells per cubic millimeter, and the sample means from any population regardless of the population distributions (by the central limit theorem).

The normal frequency curve (see Fig. 4-2a) is obtained from the binomial distribution with p equal to q equal to 0.5 when n is increased without limit. The probability function for the normal distribution is

$$P = f(x) = \frac{1}{\sqrt{(2\pi)} \, \sigma} \, e^{-(\bar{x} - x)^2/2\sigma^2} \tag{4-11}$$

where $P = f(x) =$ the probability density of size x and the other terms are as previously defined.

The normal distribution has a mean defined by Eqs. 4-1 and 4-2 and a standard deviation defined by Eq. 4-5. The mean, median, and mode are all equal (zero for the standardized variate). The standard deviation is the distance from the mean to the inflection point (1 for the standardized variate). Since the curve is symmetrical about the mean, the skewness is 0 and the kurtosis of the standardized variate is 3. The integration of the normal probability function cannot be accomplished directly; therefore, tables for the integral and the value of the function (ordinate of the frequency curve) are used (see Table 4-1). Application of the normal probability is discussed subsequently.

VII. CONFIDENCE LIMITS OR ERROR LIMITS

When the same item is measured many times, it is most unlikely that all of the measurements are exactly the same. These variations result from the inadvertent differences in conditions of measurement, errors of observation, fluctuations inherent in the process or equipment and, most of all, efforts to obtain greater precision of measurement. The amount of variation reflects the precision of the measurements--their reproducibility. The tacit assumption is often made that precise results are accurate--accuracy is the measure of correctness. The assumption is warranted if the process or instrument is frequently checked against a standard of known size. Records should be kept of these checks so that trends can be spotted, chi-square tests may be performed, and statistics can be used in designing experiments and analyzing results from these experiments.

Reference to the frequency curve in Fig. 4-2a makes evident the fact that an error limit of ϵ placed on the mean gives some area beneath the curve. This area put into a ratio with the total area beneath the curve (1 for standardized variate) gives the probability of obtaining a measurement within $\bar{x} \pm \epsilon$. The error limit is given in multiples of σ such that a determined parameter may be listed as $\bar{x} \pm k\sigma$. The limits $k\sigma$ are called the confidence limits because they express the confidence one can have that any measurement from the population will fall within the specified range.

The percent confidence limit should be carefully selected with the knowledge of what the measurement is to be used for and should be stated with the measurements. It is much better simply to give the average measurement and the standard deviation, both identified, than to give unspecified error limits. Some values of k versus the confidence limit (percent error as sometimes used) can be taken from Table 4-1; a more extensive listing can be found in most statistical tabulations.

EXAMPLE 4-4: If a measurement is 20 units with a σ of 5 units, what percent of the population should give measurements between 25 and 30 units?

k = 1 for lower measure; k = 2 for higher measure.

Area from $\bar{x}$ to $\bar{x} + \sigma$ = 34.1%   From Table 4-1.

Area from $\bar{x}$ to $\bar{x} + 2\sigma$ = 47.7%

Percent of measurements of 25 to 30 = 47.7% - 34.1% = 13.6%

= 13.4%

VIII. SKEWED DISTRIBUTIONS

Many formulas have been devised to fit skewed (nonsymmetrical) frequency curves. Some of these are the Cauchy and its modifications, chi-square, beta, gamma, skewed frequency (Slade), the dozen or so types of Pearson, the Charlier curves, and the log normal. Each has an advantage in some application; however, by far the most useful among these for air pollution analysis is the log normal. The log normal is a positively skewed (mean greater than mode, see Fig. 4-2b) frequency curve which has the same shape as a normal curve when the sizes are plotted on a logarithmic scale; that is, the logarithms of the variates are normally distributed.

TABLE 4-1

Characteristics of the Normal Probability Curve[a,b]

w	$\phi(w)$	$\int_0^w \phi(w)$	w	$\phi(w)$	$\int_0^w \phi(w)$
0.00	0.399	0.000	2.50	0.018	0.494
0.25	0.387	0.099	2.75	0.009	0.497
0.50	0.352	0.191	3.00	0.004	0.499
0.75	0.301	0.273	0.6745	0.318	0.250
1.00	0.242	0.341	1.643	0.103	0.450
1.25	0.183	0.394	1.96	0.058	0.475
1.50	0.130	0.433	2.58	0.014	0.495
1.75	0.086	0.460	2.97	0.005	0.4985
2.00	0.054	0.477	3.29	0.002	0.4995
2.25	0.032	0.488			

[a] $w = |x - \bar{x}|/\sigma$.

[b] To obtain confidence limits, use $k = w$ and confidence limit

$$= 2 \int_0^w \phi(w).$$

It should be borne in mind by the user of statistics that the parameters for the log normal are obtained by different procedures from those used with the normal, the similarity being the relations of logarithmic addition and subtraction to arithmetic addition and subtraction--addition:multiplication and subtraction:division. The mean for the log normal distribution is called the geometric mean and is represented by

$$\bar{x}_g = (x_1 \cdot x_2 \cdots x_N)^{1/N} = \left[\prod_i x_i \right]^{1/N}, \quad i = 1 \text{ to } N \quad (4\text{-}12)$$

where $\bar{x}_g$ = the geometric mean of the set of N values of x. The geometric mean is related to the arithmetic mean of the log normal set by

$$\bar{x} = \text{antiln} \, (\ln \bar{x}_g + \ln^2 \sigma_g), \quad\quad\quad (4\text{-}13)$$

where σ_g = the geometric standard deviation of x = the antilog
of the standard deviation of the logarithms of x. (see Table 8-1).
Note that ln x means the natural logarithm or logarithm to the
base e; log x means base 10; and log or logarithm without specified variate means logarithm in general without regard to base.

The confidence limits for the log normal distribution are
$\bar{x}_g \overset{x}{\div} k\sigma_g$ and the geometric standard deviation is dimensionless.

To avoid the undue influence of the sometimes very large
variates on the mean, the geometric mean should always be used
with log normal or other distributions with strong positive skewness.

IX. USE OF SPECIAL PROBABILITY GRAPHS

When the cumulative frequency of a normal population is
plotted against size, an ogee curve results (see Fig. 4-3a). By
proper change of the frequency scale, the plot will be a straight
line (see Fig. 4-3b). Figure 4-3 plots are for the data of Example 4-1 (see Table 4-2). Such graph paper is called normal proba-

TABLE 4-2

Frequency Plotting Position

Size	Cumulative frequency	$\% < \; = m/(N + 1)$
65	1	0.33
75	8	2.66
85	34	11.3
95	94	31.2
105	195	64.8
115	264	87.7
125	291	96.7
135	298	99.0
145	300	99.7

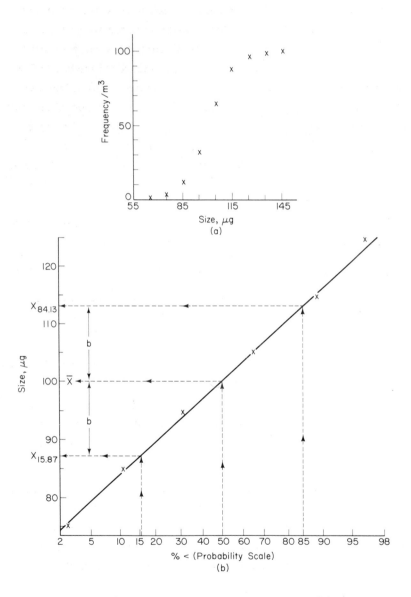

FIG. 4-3. Cumulative probability plots. (a) Ogee curve.
(b) Normal probability.

bility paper and can be purchased at book stores and stationers.
The plotting position for a given size is determined by arranging
the N items in increasing (decreasing) order by size, counting
the position (m) in the array of values, and calculating $m/(N + 1)$
as a percent for the percent less than (greater than). The 1 in
the denominator avoids obtaining a 100% value, which obviously
cannot be plotted. Care must be exercised in using the top of
the group size for percent less than and the bottom size for
percent greater than. The plot then runs uphill (downhill) to the
right, as percent less than (greater than) is the abscissa and
size the ordinate. If a straight line gives a close fit to the
plotted points, the distribution is assumed to be normal and the
mean and standard deviation are easily determined from the line.

$$\bar{x} = x_{50} \quad \text{and} \quad \sigma = x_{84.13} - x_{50} = x_{50} - x_{15.87}, \qquad (4-14)$$

where the subscripts represent the percentile values; that is,
x_{50} is the size from the 50% less than (greater than) point on
the line.

Log normal probability paper is identical with normal probability paper on the probability scale but has a logarithmic scale
for size. Of course the logs of the variate may be plotted on
normal probability paper used for the log normal distribution.

$$\bar{x}_g = x_{50} \quad \text{and} \quad \sigma_g = \frac{x_{84.13}}{x_{50}} = \frac{x_{50}}{x_{15.87}}. \qquad (4-15)$$

Other special probability graph papers have been devised.
Perhaps the only ones that merit mention here are those for extremal statistics according to Gumbel. Reference is made to his
work for any purpose involving the few extreme occurrences,
rather than using the far ends of the normal distribution.

X. REGRESSION AND CURVE FITTING

Those who desire a mechanical means of fitting a curve to
a bivariate set of data most often resort to the least squares

method (LSM). For example, to fit a straight line to a data plot
$(y = a + bx)$, the error is taken as the difference between the
measured value (y_{di}) and the calculated value (y_{ci}) of the dependent variate; then the sum of the squares of the errors,
$[\Sigma (y_{di} - a - bx_i)^2]$, is minimized by setting the partial deriva-
tives of the sums of the squares of the errors with respect to a
and b, respectively, equal to zero. Such a procedure yields two
equations to solve for parameters a and b; they are

$$Na + b\Sigma x = \Sigma y \qquad\qquad (4\text{-}16)$$

and

$$a\Sigma x + b\Sigma x^2 = \Sigma xy ,$$

where all summations are over N values of x and y. Equations
4-16 are known as the normal or normalized equations and are
used to regress y on x, a regression intended solely for using
x to estimate y, and the selection of x and y must be as the
independent and dependent variables, respectively. Of course
the LSM may be used for fitting polynomials; there are as many
equations as there are parameters of fit. All computer centers
of any consequence have a resident library program for fitting
polynomials by least squares.

There is not unanimity on the use of the LSM; however,
there does seem to be agreement that the curve should go through
the mean of the data. A method that is much easier to apply than
the LSM is the method of averages (MOA). In this method the
data are ordered by size on the x values, divided into as many
groups as there are parameters of fit (2, a and b for a line), and
the residuals (errors) in each group are set equal to zero. For a
straight line, a and b are calculated by

$$\Sigma (y_{di} - y_{ci}) = \Sigma y_{di} - \frac{N}{2} a - b \Sigma x_{di} = 0 , \qquad (4\text{-}17)$$

where the first equation is obtained by summing over i = 1 to
i = N/2 and the second by i = (N/2) + 1 to i = N. If N is odd,

simply put one more entry in one group than in the other. The method of averages gives a curve that goes through the mean of each data group as well as through the overall mean.

EXAMPLE 4-5: Fit a straight line by the LSM and the MOA to the data shown for the characteristic curve of a carbon vane pump.

Pressure (in. mercury)	Flow rate (liters/min)		
x	y	xy	x^2
3	28.2	84.6	9
6	25.0	150.0	36
9	20.0	180.0	81
12	15.6	187.2	144
15	12.4	186.0	225
18	8.0	144.0	324

LSM: $N = 6$ $\Sigma x = 63$ $\Sigma y = 109.2$
$\Sigma x^2 = 819$ $\Sigma xy = 931.8$

$6\,a + 63\,b = 109.2$
$63\,a + 819\,b = 931.8$ $a = 32.5$ $b = -1.36$

MOA:
$3\,a + 18\,b = 73.2$
$3\,a + 45\,b = 36.0$ $a = 32.7$ $b = -1.38$

The correlation coefficient as used by Pearson presumably gives a measure of the goodness of fit of a straight line used for bivariate data. The equation for the correlation coefficient (r) can be put into several forms, the most useful being

$$r = \frac{N \Sigma xy - \Sigma x \Sigma y}{[N \Sigma x^2 - (\Sigma x)^2]^{1/2} [N \Sigma y^2 - (\Sigma y)^2]^{1/2}} \qquad (4\text{-}18)$$

Considerable care must be exercised in using the correlation coefficient because it tests only for a linear fit and the value is unduly influenced by a few large values in a series. The latter case very often exists in making environmental measurements of size as a result of the low levels normally present, the inadmissibility of negative sizes, and large fluctuations in the item measured or in the method of measurement. For those

instances in which it is applicable, a value of 0 shows no correlation and +1 or -1 means perfect correlation (direct or inverse).
The correlation should always be an addition to rather than a
substitute for a plot of the data. Nonlinear and multivariate
correlations are becoming popular since the advent of the digital
computer. The techniques involved in these statistical parameters are not deemed necessary to this discussion and reference
should be made to statistics texts.

XI. STATISTICS OF RADIATION COUNTING

Radiation counting measures the number of radioactive
disintegrations in a given time or, from another viewpoint, the
average time between disintegrations. Because of the nature of
such data gathering, the average is simply the number of counts
(N) and the standard deviation is

$$\sigma = \sqrt{N} \ . \qquad\qquad (4\text{-}19)$$

Since the number of counts taken may be considered an average,
a group of measurements on a single sample should be normally
distributed.

XII. ERROR PROPAGATION

Error propagation during arithmetic operations or the taking
of logarithms is calculable by a relatively concise formula.

$$\sigma_w^{\ 2} = (\partial w/\partial x)^2 \ \sigma_x^{\ 2} + (\partial w/\partial y)^2 \ \sigma_y^{\ 2} + \cdots, \qquad (4\text{-}20)$$

where $w = \emptyset \ (x, \ y, \ \ldots)$.

EXAMPLE 4-6: A radioactive sample shows 2500 counts in 20 min
  in a counter with a background count of 400 counts in 10
  min. What is the net count rate at a 95% confidence level?

  95% Error or Confidence Limit: $\bar{x} \pm 1.96\,\sigma$

  Gross: $2500/20 \pm 1.96\,\sqrt{2500}/20 = 125 \pm 4.9$ cpm

  Background: $400/10 \pm 1.96\,\sqrt{400}/10 = 40 \pm 3.9$ cpm

  Net: $125 - 40 \pm \sqrt{(4.9^2 + 3.9^2)}$ cpm $= 85 \pm 6.26$ cpm

In other words, the error for a sum or a difference is simply the
square root of the sum of the squares of the individual errors
(with the same confidence limits).

XIII. SIGNIFICANCE TESTS

Many tests have been devised for checking the statistical
difference between statistical parameters. Despite the large
number in existence only three are briefly discussed here--chi-
square, student's t, and Chauvenet's rejection tests.

A. Chi-Square Test (χ^2)

The χ^2 test is used to test whether a group of observed
size frequencies belong to a certain distribution. The value of
χ^2 is calculated by

$$\chi^2 = \sum_i (f_{oi} - f_{ti})^2/f_{ti} \, , \tag{4-21}$$

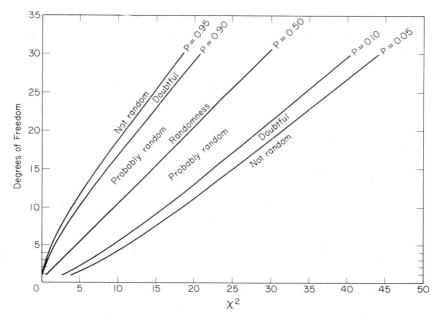

FIG. 4-4. Chi-square distribution for reliability test.

where f_{oi} = observed frequency in size interval i and f_{ti} = theoretical frequency in size interval i or frequency in the group being used for comparison. The calculated x^2 is used with a table or preferably with a graph (see Fig. 4-4) to find the probability of obtaining a x^2 of that size from random variations of the data within that distribution. A probability of 0.5 is to be expected and shows true randomness of the fluctuations. The farther the x^2 value differs from the 0.5 probability, the less likely it is that the fluctuations are only random. In order to run x^2 for radiation counting, use N_i for f_{oi} and $\bar{N}$ for f_{ti}.

EXAMPLE 4-7: If 100 measurements are made of the suspended particulate matter in a standardizing chamber with a supposed uniform concentration and the results shown in columns 1 and 2 are obtained, what does the x^2 test tell about the distribution of these measurements? Is there likely some systematic error in the method or are the fluctuations probably accidental (random)?

$f_o g^2$	$f_o g$	g	1 ($\mu g/m^3$)	2 f_o	3 f_t [a]	4 $(f_o - f_t)$	5 4^2	6 $(f_o - f_t)^2/f_t$
16	− 4	−4	38.0-39.9	1	0.5	0.5	0.25	0.50
27	− 9	−3	40.0-41.9	3	4.0	−1.0	1.00	0.25
60	−30	−2	42.0-43.9	15	16.1	−1.1	1.21	0.08
32	−32	−1	44.0-45.9	32	31.4	0.6	0.36	0.01
0	0	0	46.0-47.9	33	30.1	2.9	8.41	0.28
14	14	1	48.0-49.9	14	14.3	−0.3	0.09	0.01
4	2	2	50.0-51.9	1	3.2	−2.2	4.84	1.51
9	3	3	52.0-53.9	1	0.4	0.6	0.36	0.90
162	−56							Σ = 3.54

[a] Calculated from percent area under curve from $\bar{x}$ to $\bar{x} \pm k\sigma$.

$$\bar{x} = \text{Mid } g_o + i \, \Sigma f_o g/N = 47.0 \qquad 2\,(-56/100) = 45.88 \; \mu g/m^3$$

$$\sigma = i \left[\Sigma f_o g^2/N - (\Sigma f_o g/N)^2 \right]^{1/2} = 2 \sqrt{(1.62 - 0.31)}$$
$$= 2.29 \; \mu g/m^3$$

Eight data groups: Degrees of freedom = 8 - 1 for N - 1 for $\bar{x}$ - 1 for σ = 5

Reference to Fig. 4-4 shows that a value of $x^2 \doteq 3.54$ with 5 degrees of freedom falls very close to the probability 0.5 and the fluctuations are likely random.

B. Student's t Test

The t test is used to determine the significance of the difference between a sample mean and the estimated population mean.

$$t = \frac{|\bar{x} - \mu|}{\sigma_s / \sqrt{(N - 1)}} , \qquad\qquad (4-22)$$

where t = value of test parameter, $\bar{x}$ = sample mean, μ = population mean, σ_s = standard deviation of sample, and N = number of items in sample. There are tables of t values versus probabilities for a given N; however, a value of 2 or more may be taken as indicating a significant difference between $\bar{x}$ and μ. Specific applications of the t test are made to assign a probability that an experimental group differs from the control group or that a short-term change in a long-term measurement program is or is not probably significant.

EXAMPLE 4-8: The carbon monoxide concentration in one section of a city has been monitored over a considerable period. It has shown an approximately normal distribution with a mean of 12 ppm and a standard deviation of 5 ppm. Since a new freeway was opened nearby, the average of 10 daily samples has been 14 ppm. Is this change significant?

$t = (14 - 12)/(5/\sqrt{9}) = 2/(5/3) = 1.2$

Because 1.2 < 2 which we have set for significance, the change is not yet deemed meaningful.

C. Chauvenet's Rejection Test

There is an urge in all of us to discard a piece of data that does not agree with the other values in the set. Sometimes an aberrant measurement is the result of a real anomaly or discontinuity in the process but most often it is attributable to poor experimental technique or simply the occurrence of one of the values far out in the tail of the frequency curve. Chauvenet's criterion for rejection is

$$|x_i - \bar{x}| \geq [1 - (1/2n)]_\epsilon , \qquad\qquad (4-23)$$

where the term on the left is the absolute variation of the kth value from the mean and the right is the error associated with a $1 - (1/2n)$ confidence limit.

EXAMPLE 4-9: Five measurements are made on the concentration of hydrocarbons in an auto exhaust. The values are 425, 410, 430, 450, and 490 ppm_v. Should the 490 value be discarded?

$\bar{x} = 441$ ppm $\sigma = 31$ ppm

$x_i - \bar{x} = 490 - 441 = 49$ ppm

$[1 - (1/2n)]_e = [1 - (1/2 \times 5)]_e = 0.9$ error
$\qquad\qquad\qquad\qquad\qquad\qquad = 1.67 \, \sigma$ (from Table 4-1)

$1.67 \, \sigma = 1.67 \times 31 = 51.7 \quad |x_i - \bar{x}| < 0.9$ error

Therefore the 490 value should be retained.

The alternative to discarding a piece of data believed to be an extreme for the population is to take enough other measurements to minimize the effect of the anomalous value on the statistical parameters used to describe the data. In some cases this can be a major undertaking.

XIV. STATISTICS OF RECORDER TRACES

Measurement signals recorded on charts usually show fluctuations. Such fluctuations may be caused by noise in the instrument, variations in the measurement method, or variations in the sample property being measured (e.g., radioactive disintegration rate).

It is often desirable to determine not only the average measurement but also the standard deviation of the fluctuations. Parallel lines of constant value are drawn to enclose the fluctuation peaks and the distance between them is taken to be 4σ (sometimes 5σ), a logical assumption because of the response inertia of the instrumentation. The mean line should be halfway between the enveloping lines but is often struck to divide the total lengths of the trace lines or the areas under the peaks formed by the trace lines so that half lie on each side of the mean line.

A sufficient number of fluctuations to provide statistical accuracy should be used, a minimum of 20 peaks if possible. The size of the fluctuations in relation to the mean is important to the confidence in the measurement and is discussed in Chapter 7.

XV. SUMMARY

A rudimentary understanding of mathematical statistics is essential to presenting data properly and to understanding data that are presented. Since not all measurements are likely to be the same, it is logical that measured data often fall into a pattern. Statistics is a tool for presenting the pattern of data with a minimum amount of effort. In the statistical curve-fitting process, a set of data loses some of its information; this should always be kept in mind when transforming any data set into a form more readily observable.

A set of measurements is first divided into groups. For convenience of treatment the group sizes are usually taken equal and the number of groups from 7 to 16; about 10 is optimum. In some cases the group bounds may be set approximately logarithmic; for example, 1, 2, 5, 10, 20, 50, 100, and so on. Such is particularly appropriate for positively skewed distributions. A bar graph of frequency (frequency density for unequal group sizes) as ordinate versus size as abscissa is plotted--a frequency histogram. A smooth curve may be drawn to represent the frequency histogram--a frequency curve. The shape of the frequency curve often indicates what probability distribution function will best fit the data. The usual curves are the bell-shaped, normal curve and the positively skewed, log-normal curve.

A single size is then determined to represent the data, usually the mean or geometric mean (for the curves listed before), and is given with maximum error limits for a specified confidence level. The error limits are set from a measure of the spread (reproducibility) of the data--the standard deviation. Various

other statistical parameters are defined to give additional information about the original data set, but the principal parameters
are the mean and the standard deviation.

PROBLEMS

1. Is the arithmetic mean or the geometric mean best suited for
   averaging three dust counts of 20 million, 30 million, and
   80 million? Why?

2. An air pollution class makes 100 determinations of the concentration of suspended particulates with the following
   results: Concentration ($\mu g/m^3$), frequency; 60-64, 7;
   65-69, 24; 70-74, 40; 75-79, 21, 80-84, 8.
   Determine the mean, standard deviation, skewness, and
   kurtosis of the population.
   What are the 90% confidence limits on the mean?

3. Plot the data in problem 2 on probability paper and determine
   the mean and standard deviation from the plot.

4. Determine the geometric mean and geometric standard deviation from a log probability plot of the data in Table 2-5.
   What are the 90% confidence limits on the mean?

5. Calculate the correlation coefficient (r) for the data in Example 4-5.

6. The amount of titrant required for a colorimetric analysis is
   42 ± 0.4 ml and for the blank is 8 ± 0.1 ml at the 90% confidence level. What is the net amount of titrant required
   for the determination?

BIBLIOGRAPHY

H. Cramer, Mathematical Methods of Statistics, Princeton Univ.
   Press, Princeton, New Jersey, 1946.

E. J. Gumbel, Statistics of Extremes, Columbia Univ. Press,
   New York, 1958.

J. F. Kenney and E. S. Keeping, Mathematics of Statistics,
   Parts I and II, 2nd ed., Van Nostrand Reinhold, New York,
   1951.

Chapter 5

PROPERTIES OF GASEOUS AND PARTICULATE MATTER

This chapter briefly treats properties of gaseous and particulate matter in order to establish a rudimentary basis for air pollution analysis--laboratory analysis and evaluation. Several of the properties are quite commonly known and are listed here only in the interest of showing their importance.

I. GASEOUS MATTER

Gases and vapors are aeriform fluids whose molecules are free to act independently; therefore gases and vapors put into a closed container completely and uniformly fill the space in the container.

A. Fundamental Properties of Gases and Vapors

The ideal gas laws are usually sufficient for problems encountered in air pollution. Deviations, which are caused by attractive forces between molecules and the finite volumes occupied by the molecules (according to van der Waal's equation), are small for pressures up to several atmospheres but become excessively large as the gas nears liquefaction.

1. Ideal Gas Laws

Boyle's law ($pV = k$) and Charles' law ($V = kT$) are combined to give the familiar ideal gas law equation; that is,

$$\frac{pV}{T} = nR, \tag{5-1}$$

where p = pressure of gas, V = volume of gas, T = absolute temperature of gas, n = number of gram moles of gas, and R = universal gas constant.

EXAMPLE 5-1: Calculate R from the information that 1 gm mole of gas occupies approximately 22.4 liters at STP ($0^\circ C$ and 1 atm).

$p = 1$ atm $V = 22.4$ liters $T = 273^\circ K$ $n = 1$

$$\frac{(1 \text{ atm}) \times (22.4 \text{ liters})}{273^\circ K} = (1 \text{ gm mole}) \times R$$

$$R = 0.082 \text{ liter-atm}/^\circ K\text{-gm mole}$$

Or in cgs units:

Pressure = 1 atm = (76 cm mercury)(13.6 gm/cm^3)(981 cm/sec^2 = 101×10^6 gm/cm-sec^2 or dynes/cm^2

$$R = \frac{(22,400 \text{ cm}^3) (1.01 \times 10^6 \text{ dynes}/cm^2)}{(273^\circ K) (1 \text{ gm mole})}$$

$$= 8.3 \times 10^7 \text{ ergs}/^\circ K\text{-gm mole}$$

Perhaps the most used form of the ideal gas equation is

$$\frac{p_1 V_1}{T_1} = \frac{p_2 V_2}{T_2} ,$$

(5-2)

which is used for changing from one set of conditions to a second set.

2. Temperature, Specific Heat, and Thermal Conductivity

The temperature of a gas (solid) may be defined in statistical thermodynamics as the property of the substance caused by the average translational kinetic energy of its molecules as a result of heat agitation. A higher temperature simply means that the molecules contain more kinetic energy. At absolute zero there is no motion of the molecules. The temperature of a substance determines heat transfer to or from other substances in its vicinity.

Various temperature scales have been defined. The ones popularly used are Celsius or centigrade, and Fahrenheit and their counterparts for absolute temperatures, Kelvin and Rankine. The degree Celsius (centigrade) is defined as 1/100 of the tempera ture difference between melting ice and boiling water, and melting ice is taken as $0^\circ C$. The degree Fahrenheit is 1/180 of the difference noted, and melting ice is $32^{\circ}F$. The conversion from $^{\circ}C$ to $^{\circ}F$, or vice versa, is simple and is made by

$$^{\circ}F = 9/5 \ ^{\circ}C + 32. \tag{5-3}$$

To obtain absolute temperature in $^{\circ}K$, add 273.16 to $^{\circ}C$, and $^{\circ}R = 459.69 + \ ^{\circ}F$. The numbers added are usually rounded to 273 and 460.

The term specific heat when applied to a substance denotes the thermal capacity of the material or the amount of heat required to cause a unit change in the temperature of a unit mass of the substance. The units for the specific heat are $cal/gm-^{\circ}C$ and $Btu/lb-^{\circ}F$. The actual specific heat of a gas varies slightly with the temperature and the pressure. Within the range of temperatures and pressures of usual interest, the specific heat of air at a constant pressure (c_p) may be taken as $0.24 \ cal/gm-^{\circ}C$ or $0.24 \ Btu/lb-^{\circ}F$, and at a constant volume (c_v) as $0.17 \ cal/gm-^{\circ}C$. The relationship of specific heats to the gas constant is represented by

$$R = J \ (c_p - c_v), \tag{5-4}$$

where $J = 778$ ft-lb/Btu. The ratio of c_p to c_v is often denoted by γ or k and is about 1.4 for air. Values of c_p and γ for other common gases are approximately 0.21 and 1.27 for carbon dioxide, 0.22 and 1.4 for oxygen, and 0.25 and 1.4 for nitrogen.

Heat is conducted through a medium if the kinetic energy is passed from one molecule to another while the molecules retain their relative positions. For gaseous and vaporous materials, conduction occurs only in the boundary layers, while the much faster transfer of convection prevails elsewhere. Thermal conductivity (k) is the rate constant for heat conduction. The k value for air is about $0.013 \ Btu-ft/hr-ft^2-^{\circ}F$ at NTP and increases to 0.021 at $500^{\circ}F$. These values compare with the asbestos insulation k value of 0.095 and the copper k value of 221.

3. Viscosity

The viscosity of a gas (or fluid) is its resistance to flow. This is a measure of the shearing stress when one part of the gas is moved relative to an adjacent part. The dynamic viscosity (μ) is the force per unit area required to maintain a unit velocity and is given in poises. A poise is 1 dyne sec/cm^2 or 1 gm/sec-cm. The viscosity is practically independent of pressure (up to several atmospheres) but varies with the absolute temperature to the 3/2 power according to Sutherland. The dynamic viscosity of air at NTP is approximately 1.8×10^{-4} poises (see Table 1-2). The kinematic viscosity (ν) is the dynamic viscosity divided by the density (μ/ρ) and has units of stokes or square centimeters per second. It is important to note that the viscosity of a gas increases with temperature.

4. Partial Pressures

In a mixture of gases, each gas exerts a pressure in direct proportion to its abundance in the mixture; this pressure is termed the partial pressure for a particular gas. The total pressure of a mixture of gases equals the sum of the partial pressures of all its constituents.

EXAMPLE 5-2: Calculate the partial pressures in the alveolar sacs of the lung if the percentages by volume are the following: nitrogen (+argon), 75.8; oxygen, 13.3; carbon dioxide, 5.7; water, 5.2.

Total pressure = 1 atm = 760 mm mercury

Partial pressures

Nitrogen (+argon)	0.758 X 760 =	576 mm mercury
Oxygen	0.133 X 760 =	101 mm mercury
Carbon dioxide	0.057 X 760 =	43 mm mercury
Water	0.052 X 760 =	40 mm mercury
	Total	760 mm mercury

B. Gaseous Diffusion

Gaseous diffusion at the molecular level follows Fick's law that is defined for rectangular coordinates by

$$\frac{dc}{dt} = -\not{D}\,\nabla^2 c = -\not{D}\left[\frac{\partial^2 c}{\partial x^2} + \frac{\partial^2 c}{\partial y^2} + \frac{\partial^2 c}{\partial z^2}\right], \qquad (5-5)$$

where c = concentration of diffusing gas, t = time, and $\not{D}$ = diffusivity coefficient of gas (0.139 cm^2/sec for carbon dioxide in air at STP).

The rate of diffusion of a gas through a porous material is functionally related to the molecular weight of the gas by

$$\not{D}_1/\not{D}_2 = \sqrt{(MW_2/MW_1)}, \qquad (5-6)$$

C. Adiabatic Expansion

An adiabatic expansion of a gas is one in which heat is not allowed to enter or to leave the expanding gas. This hypothetical situation is practically true for relatively rapid expansion as long as the gas does not reach saturation point for its moisture content. The concept is quite important in atmospheric stability considerations (see Chapter 3). The isothermal condition of Boyle's law is modified for adiabatic conditions to

$$pV^\gamma = constant, \qquad (5-7)$$

where γ = ratio of specific heats (c_p/c_v). The expanding gas is cooled because of the work that it does in moving its surroundings and the work that it does in lessening the cohesive forces on its own molecules (Joule-Thompson effect).

D. Gas Flow in Ducts and Conduits

Confined gas flow must be treated as compressible flow for those cases in which the density changes appreciably (for example, $\Delta\rho/\rho \approx 5\%$). When rather simple equations for incompressible fluid flow are not adequate, they must be replaced by thermodynamically derived equations. Density and velocity along a uniform, horizontal conduit section are no longer constant.

1. Continuity

The mass flow rate (Q_m) must be the same at any section;

$$Q_m = \rho A v, \tag{5-8}$$

where Q_m = mass flow rate, ρ = density of gas, A = cross-sectional area of flow, and v = average velocity of flow across A.

2. Energy Equation

The energy equation for compressible fluid flow in a conduit (one-dimensional) is

$$JH_1 + \frac{v_1^2}{2g} + Jq = JH_2 + \frac{v_2^2}{2g}, \tag{5-9}$$

where J = Joule's equivalent = 778 ft lb/Btu, H_i = enthalpy at ith section = $E + pV/J$, E = internal energy, v_i = velocity at ith section, and q = change in heat content from section 1 to section 2.

By use of the energy equation and the assumptions of adiabatic flow (q = 0) and frictionless flow (isentropic conditions), an energy equation for such flow is derived as

$$\frac{v_2^2 - v_1^2}{2g} = \frac{\gamma}{\gamma - 1} RT_1 (1 - T_2/T_1), \tag{5-10}$$

where all terms are as previously defined.

3. Head Loss in Conduit Flow

For most calculations of head loss in ducts (all of those with no practical change in temperature), the head loss may be determined from

$$\frac{\Delta p}{L} = \frac{f'RTQ_m^2}{40\rho D^5}, \tag{5-11}$$

where $\Delta p/L$ = pressure drop in length L (lb/ft^2), f' = friction factor (for rough ducts, f' = 0.052 when N_{Re} = 3000 and 0.032

when $N_{re} \geq 10^6$ and for smooth ducts, $f' = 0.050$ when $N_{Re} = 3000$ and 0.013 for $N_{Re} = 10^7$), R = gas constant (ft-lb/lb-°R), T = absolute temperature of gas (°R), Q_m = mass flow rate (lb/sec), p = pressure (mean in lb/ft^2), and D = duct diameter (equivalent for nonround = $\sqrt{area/0.785}$, in ft).

E. Evaporation and Vapor Pressure

Evaporation is the escape of molecules from the main body of a liquid into the air or gas contacting the liquid. Sublimation is the equivalent process for a solid-to-gas conversion. The escaping molecules must have sufficient energy to escape the liquid surface; they must overcome surface tension and intermolecular forces. This energy is supplied to the molecules by heat from the surroundings or by a source applied to the liquid. Thus evaporation results in cooling the surrounding gas.

After the molecules leave the liquid, they exist in the gas as a vapor and exert pressure there. When the number of molecules leaving the liquid or solid in a given time equals the number of like molecules returning to the substance by condensation, an equilibrium exists and the partial pressure of the vapor phase is known as the saturated vapor pressure (or vapor pressure). Vapor pressure varies with the temperature approximately as

$$p_v = \exp (f + g/RT), \tag{5-12}$$

where p_v = vapor pressure, f and g = parameters of fit, and other terms are as previously defined (see Table 1-2). The differential form of the equation is known as the Clausius-Clapeyron equation. It is

$$\frac{d(\ln p_v)}{dT} = \frac{L}{RT^2}, \tag{5-13}$$

where L = the latent heat of evaporation (cal/mole, for water L = 9705, for ammonia L = 5580, for mercury L = 14,275, and for sodium L = 23,675). Note that L = -g in Eq. 5-12. When the vapor pressure reaches the total pressure on the liquid, the liquid is at its boiling point.

Vapor pressure of a solvent is changed by the presence of a solute according to Raoult's law, or

$$\frac{p_0 - p_s}{p_0} = \frac{n_s}{n_0 + n_s} , \qquad (5-14)$$

where p_0 = vapor pressure of pure solvent, p_s = vapor pressure of solvent above solution, n_s = gm moles of solute, and n_0 = gm moles of solvent.

F. Solubility or Absorption

Gases dissolve to some extent in liquids that they contact. This forms the basis of human respiration. It is also very important in air sampling and air pollution control. Many slightly soluble gases follow Henry's law which states that the amount of gas dissolved in a liquid is directly proportional to the partial pressure of the gas in contact with the liquid, or

$$p = Hx , \qquad (5-15)$$

where p = partial pressure of gas (mm Hg), x = mole fraction of gas in liquid, and H = Henry's constant (mm Hg). H varies with the gas, the solvent, and the temperature. A few values of H for water at $20^{\circ}C$ $(10^{\circ}-30^{\circ})$ are: carbon dioxide, 0.108 (0.079-0.139); oxygen, 2.95 (2.48-3.52); nitrogen, 5.75 (4.87-6.68); hydrogen, 5.20 (4.82-5.51). In addition to holding for slightly soluble gases, the equation applies to the low concentrations of air pollutant gases usually found in the atmosphere, regardless of their solubilities. Note that higher temperatures require higher pressures for dissolving the same mole fraction.

G. Adsorption

Gaseous (or liquid or dissolved) molecules that adhere to the surface of a solid by physical or chemical attraction are said to be adsorbed. The surface is the adsorbent and the material adsorbed is the adsorbate. The amount of material adsorbed depends upon the number of available "active" sites on the adsorbent. Although there are several materials that adsorb, including activated carbon or charcoal, silica gel, alumina, clays, and powders, the only practical adsorbent for most air pollution work is activated carbon. Activated carbon is the only nonpolar adsorbent of note and therefore the only one that can be used satisfactorily in the presence of water vapor. Silica gel is often used for drying air streams.

The equation for the relative amount of material that adsorbs at a given temperature is called an adsorption isotherm. Freundlich's adsorption isotherm is a popularly used empirical equation. It is

$$x/M = kp^{1/n}, \tag{5-16}$$

where x = mass (gm) of gas adsorbed on M grams of adsorbent, p = pressure of gas in bulk phase, and k and n = parameters of fit or constants for a particular case.

The Langmuir adsorption isotherm is derived by defining the rates of adsorption and desorption in terms of the proportion of available sites occupied. The steady state has adsorption equal to desorption.

$$k_a p (1 - \phi) = k_d \phi, \tag{5-17}$$

where ϕ = fraction of adsorption sites occupied, k_a = rate constant of adsorption, and k_d = rate constant of desorption. If $k_a/k_d = k$, $\phi = (kp)/(1 + kp)$.

In the latter equation, ϕ is proportional to x/M and, by alternately considering the cases for $kp \ll 1$ and $kp \gg 1$, it is seen that the power of p goes from about 1 to near 0, which agrees with the Freundlich equation.

Calculations for adsorption in a liquid are made by substituting concentration (c) for pressure (p). Further, the tendency of a solute to concentrate at the surface of a liquid (which lowers surface tension) is termed adsorption and may be calculated from the Gibbs adsorption isotherm.

$$\Gamma = -\frac{1}{RT} \times \frac{d\sigma}{d(\ln c)} \quad , \tag{5-18}$$

where Γ = number of excess (over c) gram moles of solute per square centimeter, R = universal gas constant, T = absolute temperature, σ = surface tension, and c = concentration of solute in the bulk.

Activated carbon is made from hardwood, coconut hulls, animal bones, peach pits (which release cyanide), and many other materials. It has an extremely large surface area because of pores running through the particles. The total area of a 1/16-in. particle of activated carbon is said to be many square feet.

Physical adsorption cannot be used for the true gases (critical temperatures $<\sim -50^{\circ}C$ and b.p. $< -150^{\circ}C$), such as hydrogen, nitrogen, oxygen, carbon monoxide, and methane. Materials having critical temperatures between $0^{\circ}C$ and $150^{\circ}C$ and boiling points between $-100^{\circ}C$ and $0^{\circ}C$ can be adsorbed by a long contact time (enhanced by cooling). This group includes hydrogen sulfide, ammonia, hydrogen chloride, formaldehyde, and ethane. The vapors with boiling points higher than $0^{\circ}C$ are easily adsorbed. The amount of adsorbed material (percent by weight of adsorbent) retained after exposure to saturated air and flushing with clean air is the retentivity of the adsorbent (see Table 5-1).

H. Gaseous Scatter and Absorption of Photons

Photons may be scattered by or absorbed by gaseous molecules and atoms. The portion of the spectrum of most interest in air pollution is that with wavelengths (λ) from about 300 to

TABLE 5-1

Retentivities of Vapors by Activated Carbon[a]

Compound	Formula	Retentivity x/M (%)	Description
Acetaldehyde	CH_3CHO	7	Apple ripener
Acrolein	CH_2CHCHO	15	Hot fat odor
Amyl acetate	$CH_3COOC_5H_{11}$	35	Solvent for lacquers
Butyric acid	C_3H_7COOH	35	Dirty feet odor
Carbon tetrachloride	CCl_4	45	Solvent; standard for retentivity specifications
Ethyl acetate	$CH_3COOC_2H_5$	20	Solvent for lacquers
Ethyl mercaptan	C_2H_5SH	25	Garlic, onion, sewer
Eucalyptole	$C_{10}H_{18}O$	20	Inhalation medicant
Formaldehyde	$HCHO$	3	Disinfectant
Methyl chloride	CH_3Cl	5	Grain fumigant (formerly refrigerant)
Phenol (carbolic acid)	C_6H_5OH	30	Hospital odor
Putrescine	$NH_2C_4H_8NH_2$	25	Decaying flesh
Skatole	$C_6H_4C_2H(CH_3)NH$	25	Excreta
Sulfur dioxide	SO_2	10	Burning coal odor
Toluene	$C_6H_5CH_3$	30	Solvent, TNT manufacturing
Valeric acid	C_4H_9COOH	7	Body odor

[a]Adapted from: A. Turk, et al., Am. Industrial Hyg. Assoc. Quart., 13, 23, 1952 and others.

700 nm, the visible region plus a small amount of the UV region.
Solar radiation delivers significant energy to the lower elevations
of the atmosphere only within this region. The interactions of
other wavelengths with the air and airborne gaseous materials
are of concern in radar, radio, and microwave transmissions, in
high-energy photochemistry in the upper atmosphere, and in
spectral analysis of the airborne materials.

Rayleigh (1890) predicted that gases should scatter light
because on the microscopic scale gas is made up of small particles (molecules) separated by space with a refractive index
of 1 and is not homogeneous. He postulated that the scattering
coefficient (b_r, see Chapter 10, Section III, A, 1) is related to
the index of refraction (m) and the wavelength (λ) according to

$$b_r = km^2/\lambda^4. \qquad\qquad (5\text{-}19)$$

Rayleigh scattering of photons by interaction with atoms
does not change the internal energy of the atom. This type of
scattering is a major scattering process only for particles with
diameters much less than the wavelength scattered; that is, for
visible wavelengths molecular particles are the maximum size.
Rayleigh scattering is approximately 16 times as effective for
blue light as for red $[\lambda_b \approx (1/2)\lambda_r; 1/(1/2)^4 = 16]$. As a result,
sky light scattered into the eyes is predominately blue, whereas
a view into the light source has much of the blue scattered out
and is mostly red, as in sunsets.

Einstein developed the mathematical qualitative and
quantitative explanation for Rayleigh scattering. His basis was
that the molecular density of small volumes varied because of
Brownian motion and this variation caused the optical densities
to fluctuate. The equation he derived for solutions of liquids
or gases is

$$b_r = \frac{32\pi^3 m_0^2 RTMC}{3\, N_a \lambda^4} \left[\frac{dm}{dC}\right]_{pT}^2, \qquad (5\text{-}20)$$

where b_r = Rayleigh scatter coefficient, m_0 = index of refraction
of solvent (air), m = index of refraction of solution, R = universal
gas constant, T = absolute temperature, M = molecular weight of
solute, C = concentration (gm/liter), N_a = Avogadro's number,
and λ = wavelength of scattered photon. Further treatment of
Rayleigh scattering may be found in van de Hulst as cited in the
bibliography.

The absorption of electromagnetic radiations by gases is
spectral in nature. The molecules (atoms) can absorb only those
energies that raise the internal energy to a level permitted by
the principles of quantum mechanics. The absorption of note
for gaseous materials normally found in air and within the 300-
to 700-nm wavelengths is the secondary band for oxygen, 687-
692 nm. Other gaseous materials which are found in the air and
enter into absorption of visible light are ozone, nitrogen dioxide,
sulfur dioxide, nitric acid, and many organics. The important
nonabsorbers include nitrogen, water, carbon dioxide, carbon
monoxide, sulfur trioxide, sulfuric acid, hydrocarbons, alcohols,
and organic acids.

Application of these principles is discussed in Chapter 7
and in Chapter 10.

II. AEROSOLS

Aerosols are air or gaseous suspensions of solid or liquid
particles. In order for particles to remain suspended for a sin-
nificant amount of time, the settling velocity of the particle
must be small and the diameter usually less than 50 μm. The
small size of the particles gives them a large surface area-to-
mass ratio, a factor that accounts for many of their chemical and
physical properties.

A. Condensation on Particles

Particles in the size range of about $0.001-0.1$ µm serve as condensation nuclei. The degree of supersaturation must be approximately 4.2 times as large for self-nucleation as for condensation on nuclei. Large numbers of particles which can serve as condensation nuclei are produced by smokes, dusty processes, and resuspension. The particles have a rather long lifetime in the atmosphere and the number present averages about $1000/cm^3$ in the cleanest ocean and mountain air to about $150,000/cm^3$ in cities but may be as high as 20 times these values.

The work of forming particles for self-nucleation is expressed as

$$W = \frac{16\pi\sigma^3 V}{3(\bar{K} T \ln p'/p)^2} ,$$

(5-21)

where W = work of formation, σ = surface tension of liquid, V = volume per molecule of droplet, $\bar{K}$ = Boltzmann's constant, and p'/p = degree of supersaturation.

The work is reduced when nuclei are present by a factor of

$$\frac{(2 + \cos\bar{\Phi})(1 - \cos\bar{\Phi})^2}{4} ,$$

(5-22)

where $\bar{\Phi}$ = angle of contact between the growing aggregate of water molecules and the surface of the particle.

Condensation in pores of particles under a vacuum produces anomalous water or polywater with a density of 1.3. Such water exerts a much greater adhesive force than the simple surface tension of water. Fish uses structured capillary water to explain the forces he observed between particles and surfaces, forces that were several times the predicted surface tension forces.

B. Evaporation of Droplets

Droplets have higher vapor pressures than the same liquids with plane surfaces because of the difference in surface tension forces per unit volume.

$$\ln \frac{p_v{}'}{p_v} = \frac{1}{\rho RT} (2\sigma/r - f),$$ (5-23)

where $p_v{}'$ = vapor pressure over curved surface, p_v = vapor pressure over plane surface, σ = surface tension, ρ = density of liquid, r = radius of curvature of surface, R = universal gas constant, T = absolute temperature, and f = 0 for uncharged particles = $\varepsilon^2/8\pi r^4$ for particles with a charge of ε/cm^2.

The lifetime of droplets in a saturated air is calculated by

$$t = \frac{r^3}{6\not{D}\sigma p_v} (\rho RT/M)^2,$$ (5-24)

where t = lifetime of droplet (sec), $\not{D}$ = diffusivity of vapor in air = 0.220 cm^2/sec at $0°C$, and M = molecular weight of liquid = 18 for water.

A 2-μm particle (radius = 1 μm) of water disappears in 2.6 sec in a saturated environment at $0°C$. Large particles grow at the expense of small particles.

C. Charge on Aerosols

Aerosols may take a charge that affects their behavior. The famous Millikan oil drop experiment measures the magnitude of the elemental charge by the velocity of small, charged particles in an electric field. Aerosols may be charged by collision with ions produced by nuclear radiations, flame, or other energy sources. Particles become charged when liquid is dispersed in a gas (shower), water vapor expands suddenly into air through narrow opening, and other processes that cause an increase in specific surface.

The rate of change in particle charge by collision with diffusing ions has been formulated as

$$\frac{dn}{dt} = \pi r^2 wN e^{-ne^2/rkT}, \quad d > \lambda$$ (5-25)

where n = number of elemental charges, t = time (sec), w = mean thermal ion velocity $\approx \sqrt{T}$ (cm/sec), ε = elemental charge =

4.8×10^{-10} esu, r = particle radius (cm), N = ion concentration (Nϵ = space charge) (cm^{-3}), $\bar{K}$ = Boltzmann constant (erg/^{O}K), and T = absolute temperature (^{O}K).

For an initially uncharged particle,

$$n = \frac{rkT}{\epsilon^2} \ln (1 + \frac{\epsilon^2 \text{rmwN}}{kT}) ,$$ (5-26)

and for small t and N,

$$n = \pi Nwr^2 t .$$ (5-27)

Pauthenier derived a relation for the charging of particles by ion bombardment in an electric field.

$$n\epsilon = (1 + 2 \frac{\xi - 1}{\xi + 2}) r^2 E \frac{1}{1 + (1/\pi N\epsilon ut)} ,$$ (5-28)

where ξ = dielectric constant for particle, E = electric field strength (V/cm), and u = mobility of the charge carriers (cm^2/V-sec) Nϵu = j/E, where j = current density (A/cm^2)

Whitby formulated a relation for mean free path much greater than diameter as

$$\frac{dn}{dt} = \frac{\pi r^2 j}{\epsilon}$$ (5-29)

or

$$n\epsilon = \pi r^2 jt.$$ (5-30)

D. Photon Scattering and Absorption by Aerosols

Mie scattering of photons is the major scattering process for particles with diameters much greater than those for the molecules considered for Rayleigh scattering. Mie particles have diameters of the same order of size as the wavelength. The derivation of Mie (Debye) mathematics for light scattering by a single particle involves Maxwell's equations to describe the oscillations exterior to the particle. The Mie scattering for spherical water particles of various diameters is shown in Fig. 5-1 The forward scattering is greater for large values of α ($\alpha = 2\pi r/\lambda$), that is, for large radius particles or short wavelengths. A small

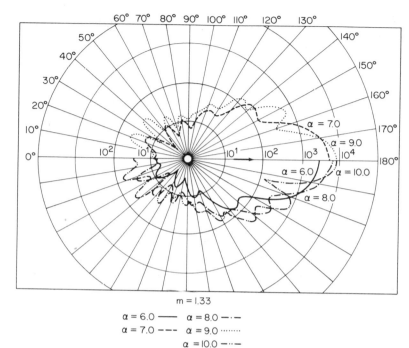

FIG. 5-1. Mie scattering by spherical water droplets.
From H. Neuberger, H. Panofsky, and Z. Sekera, "Physics of
the Atmosphere," <u>Air Pollution Handbook</u>, P. Magill, F. Holden,
and C. Ackley, Eds., Section 4, McGraw-Hill, New York, 1956.

index of refraction (m) favors forward scattering and a large m
favors backscatter; for opaque materials the asymmetry is toward
backscatter.

Mie scattering has application in particle size analysis by
corona effects (Chapter 8, Section II, D, 4) and in visibility
considerations (Chapter 10, Section III, A, 1).

E. Melting Point, Boiling Point, and Thermal Conductivity

The temperature at which solid particles become liquid,
the melting point, and the temperature that causes a vapor pressure of 1 atm for liquid particles, the boiling point, are important
in particulate analyses and sometimes in particulate air cleaning.

Evaporation of solid particles or sublimation can occur but is not of major significance for most of the particles of concern in aerosol pollution.

The thermal conductivity of particles gives a measure of heat conduction in the particles and does not depend upon particle size or shape, except indirectly on the particle area. The factors directly involved are shown in

$$\frac{dQ}{dt} = kA\frac{dT}{dx} \, , \qquad\qquad (5-31)$$

where dQ/dt = rate of heat transfer through area A perpendicular to direction x (Btu/hr), k = thermal conductivity (Btu-ft/hr-ft^2-$^{\circ}$F), dT/dx = temperature gradient ($^{\circ}$F/ft), and A = area of heat transfer (ft^2). The thermal conductivity of a particle is important to its mechanical behavior, especially to thermophoresis.

F. Solubility

The solubility of a substance is related to its particle size. Solids have a surface tension which provides a driving force to put small particles into solution. The heats of solution can be measured calorimetrically for large salt crystals and for finely divided salt. The difference between the two heats as the particle size decreases can be taken as a measure of the surface tension [$\Delta H = A\,(\sigma - T\,d\sigma/dt)$]. The surface tensions have been found to be about the same for the solid as for the liquid state of the same substances. Covalent compounds have weak surface tensions (10-100 dynes/cm), while ionic compounds and metals have strong surface tensions (100-1000 dynes/cm).

G. Aerosol Mechanics

The discussion of aerosol mechanics presented here is quite brief and treats the subject only in relation to practical applications.

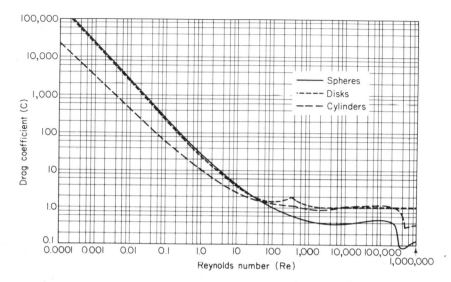

FIG. 5-2. Drag coefficients versus Reynolds numbers.
From C. E. Lapple, "Separation of Dusts and Mists," Chemical
Engineers' Handbook, J. H. Perry, Ed., 3rd ed., McGraw-Hill,
New York, 1950.

1. Resistance to Motion

A particle encounters resistance in moving through a gaseous
medium. This resistance is formulated as

$$R = CA\rho_a u^2/2 \ , \tag{5-32}$$

where R = resisting force, A = projected area of particle, ρ_a =
density of gaseous medium, u = velocity of particle relative to
medium, and C = drag coefficient. C varies with the Reynolds
number ($N_{Re} = \rho_a du/\mu$) and the shape of the particle. Lapple
and Sheppard divided the curve of C versus N_{Re} for spheres into
three zones (see Fig. 5-2), each of which may be well represented
by a straight line on the log log plot.

The low N_{Re} (<3) region is termed the streamline or Stokes
region and has

$$C = 24/N_{Re} \qquad \text{(slope = -1)}. \tag{5-33}$$

For small spheres $R = 3\pi\mu du$. This expression recurs repeatedly in aerosol mechanics. The intermediate region ($3 \le N_{Re} \le 10^3$) is represented by

$$C = 14/\sqrt{N_{Re}} \qquad (\text{slope} = -1/2).$$ \hfill (5-34)

In the turbulent region ($N_{Re} > 10^3$), the drag is independent of N_{Re} and

$$C = 0.44 \qquad (\text{slope} = 0).$$ \hfill (5-35)

2. Brownian Motion

Small particles (less than a few micrometers diameter) in a fluid suspension are differentially bombarded by the molecules of the surrounding medium. The bombardment causes actual movement of the particles--Brownian motion. Although the movements are random in direction (neglecting settling) and should average to zero total displacement over a considerable time, the root-mean-square of the displacements is not zero. Einstein's equation for calculating Brownian movement is

$$\overline{s^2} = 2\emptyset t,$$ \hfill (5-36)

where $\overline{s^2}$ = mean square displacement of particle in any direction, t = finite time for displacement, and $\emptyset$ = diffusivity coefficient for particle (see Table 5-2).

$$\emptyset = kTM,$$ \hfill (5-37)

where k = Boltzmann's constant, T = absolute temperature, and M = mobility of particle = Stokes settling velocity divided by the product of particle mass and gravity.

If the value of $\emptyset$ for spherical particles ($\emptyset = RT/3\pi\mu d$) is substituted into Eq. 5-36, the more useful form is obtained.

$$s = \left\lceil \frac{RT}{N_a} \frac{t}{3\pi\mu d} \right\rceil^{1/2},$$ \hfill (5-38)

where s = magnitude of movement of the particle in time t and N_a = Avogadro's number.

TABLE 5-2

Some Mechanical Properties of Aerosols[a]

Diameter (μm)	Diffusivity (cm^2/sec)	Mean Brownian velocity (cm/sec)	Mean free path λ (nm)	Brownian movement distance (cm)
0.002	1.28×10^{-2}	4965	65.9	1.31×10^{-6}
0.004	3.23×10^{-3}	1760	46.8	2.62×10^{-6}
0.01	5.24×10^{-4}	444	30.0	6.63×10^{-6}
0.02	1.35×10^{-4}	157	22.0	1.37×10^{-5}
0.04	3.59×10^{-5}	55.5	16.4	2.91×10^{-5}
0.1	6.82×10^{-6}	14.0	12.4	8.64×10^{-5}
0.2	2.21×10^{-6}	4.96	11.3	2.24×10^{-4}
0.4	8.32×10^{-7}	1.76	12.1	6.73×10^{-4}
1	2.74×10^{-7}	0.444	15.3	3.47×10^{-3}
2	1.27×10^{-7}	0.157	20.6	1.28×10^{-2}
4	6.10×10^{-8}	0.0555	28.0	4.93×10^{-2}
10	2.38×10^{-8}	0.0140	43.2	3.02×10^{-1}
20	1.38×10^{-8}	0.00496	60.8	1.21×10^{0}

For $\rho_p \neq 1$, multiply columnar value by ρ_p^x,
 where x =   0   -1/2   1/2   1

[a]Adapted from: N. A. Fuchs, The Mechanics of Aerosols, Pergamon Press, New York, 1964.

3. Coagulation

Small particles moving about by Brownian motion often collide and stay together to form larger particles. The rate of coagulation for a monodisperse (all particles of same size) aerosol (neglecting settling) is

$$-\frac{dn}{dt} = HC_c n^2, \qquad\qquad (5\text{-}39)$$

where n = concentration of particles (cm^{-3}), t = time, C_c = Cunningham's correction coefficient (see Eq. 5-43), and H = coagulation constant = $4\,kT/3\mu$. Measured values of H range from about 2 to 10 times the theoretical value of 3×10^{-10} cm^3/sec at $20^{\circ}C$ in air. When Eq. 5-39 is integrated over a period t, it gives

$$1/n - 1/n_0 = HC_c t, \qquad\qquad (5\text{-}40)$$

where n_0 = initial concentration of particles (n for $t = 0$).

4. Sedimentation

If the resistance a particle encounters in falling is equated to the force on the particle, a terminal settling velocity or steady-state equation is obtained.

$$R = F = V\rho_b g = \pi/6 \ \ d^3(\rho_p - \rho_a) g \qquad\qquad (5\text{-}41)$$

for spherical particles, where R = resistance to motion (see Eq. 5-32), F = gravitational force on particle, V = volume of particle, ρ_b = buoyant density of particle, ρ_p = density of particle, ρ_a = density of medium (air), d = particle diameter, and g = acceleration of gravity. Substitution of R in the Stokes' region $(R = 3\pi\mu d)$ yields the equation for the settling velocity of a small sphere; namely,

$$u_t = \rho_p g d^2/18\mu, \qquad\qquad (5\text{-}42)$$

neglecting buoyancy, where u_t = terminal settling velocity of particle. For air at NTP and d in micrometers, $u_t = 0.003\,\rho_p d^2$.

When the particles are small with respect to the mean free path of the air molecules (particles less than a few micrometers), the air can no longer be considered continuous and a correction must be applied to u_t. Although treatments by Oseen and Sutherland are considered more accurate, multiplication of u_t by Cunningham's correction for particle slip approximates actual conditions closely enough for practicality.

$$C_c = 1 + 1.7 \, \lambda/d, \qquad (5-43)$$

where C_c = Cunningham's correction factor and λ = mean free path of air molecules. For air at NTP, $\lambda = 0.06 \, \mu m$ (see Table 1-2) and the correction for a 2-μm particle is approximately 5%.

In the intermediate region of settling ($3 \leq N_{Re} \leq 10^3$) where the drag is approximately $14/\!/N_{Re}$, the velocity in air at NTP is

$$u_t = 0.34 \, \rho_p^{2/3} d, \qquad (5-44)$$

where d is in micrometers. For turbulent settling ($N_{Re} > 10^3$) where the drag can be approximated by 0.44, the equation for air at NTP and d in micrometers is

$$u_t = 16 \, \rho_p^{1/2} d^{1/2}, \qquad (5-45)$$

For unit density spheres, streamline settling applies to particles with diameters less than 115 μm, turbulent settling to those greater than 2130 μm, and intermediate settling between these limits. Figure 5-3 shows the Stokes (streamline) settling velocities.

EXAMPLE 5-3: What is the theoretical settling velocity for a 10-μm particle of ferric oxide?

$\rho_p = 5.24 \, gm/cm^3$ (from Handbook of Physics and Chemistry)

$u_t = 0.003 \, (5.24) \, (10) = 0.157 \, cm/sec$

C_c is insignificant and the N_{Re} of 1.07×10^{-4} is well within the Stokes' region.

CHARACTERISTICS OF PARTICLES AND PARTICLE DISPERSOIDS

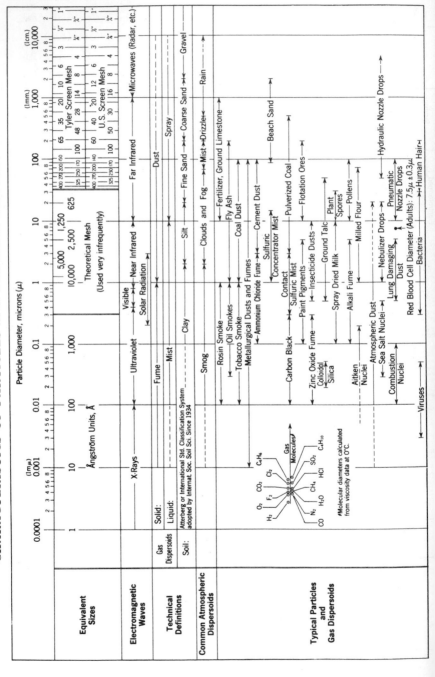

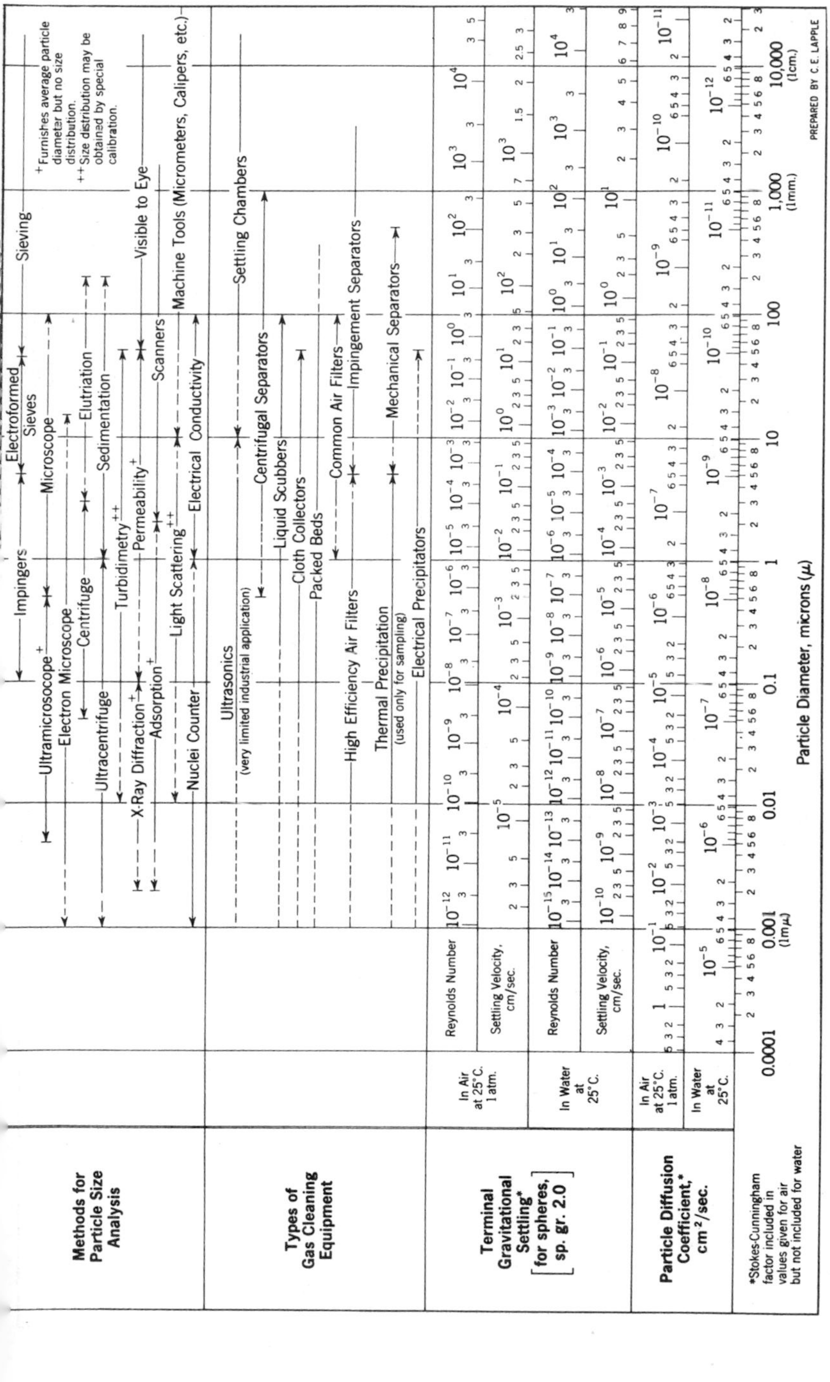

FIG. 5-3. Characteristics of particles and particle dispersoids.
Reprinted by permission of the Stanford Research Institute.

If small particles are settling in a stirred (turbulent) atmosphere, the rate of removal of particles is proportional to the number of suspended particles, which is the differential form for a first-order reaction and upon integration gives

$$n = n_0 e^{-u_t t/H}, \qquad\qquad (5-46)$$

where n = number of particles in the air or concentration of particles, n_0 = number of particles at time t = 0, H= height of fall of particles, and u_t = terminal settling velocity in quiescent conditions.

Acceleration of particles to the terminal settling velocities occurs so rapidly that it is usually neglected. For the normal sizes dealt with, the distances for the accelerations to occur are in millimeters or at most of the order of a centimeter.

Some limitations on the sedimentation equations are caused by the tendency for an envelop of air to move with individual particles. When the aerosol is in a cloud with these envelops filling all of the space, the air passes around the entire cloud and the cloud settles much faster than the discrete particles of the same size because the ratio of the area of drag to the weight of the cloud is much less than it was for the individual particles. If the cloud is in a chamber in which there is no free space around the cloud, it will be hindered in its fall and slowed to a large extent. A particle whose envelop of air intersects a wall settles more slowly than a similar particle away from the wall.

Sedimentation can be used to measure particle size, sort particles by aerodynamic size, determine density, or any parameter in the equations for settling.

5. Centrifugation

Centrifugal forces can be developed to many times gravity and move a particle through the air much faster than its settling velocity. The centrifugal force put on particles by moving an aerosol in a circular path is termed the separation factor;

$$s.f. = v_t^2/gR \, , \qquad\qquad (5-47)$$

where s.f. = separation factor (gravities, or g's), g = acceleration of gravity, v_t = tangential velocity, and R = radius of rotation.

EXAMPLE 5-4: What is the separation factor for a centrifuge with a 4-in. radius of rotation and a head speed of 4800 rpm?

$$v_t = 4800/60 \text{ rps} \times 2\pi \, 4/12 \text{ ft} = 220 \text{ ft/sec}$$
$$s.f. = (220)^2/[32.2 \, (4/12)] = 4820 \text{ g's}$$

6. Impaction

An aerosol moving toward an obstacle may impact particles on the obstacle. Because the inertia of the particles tends to cause them to move in straight lines, they do not follow the streamlines of the air around an obstacle but cross streamlines and impact on the deflecting surface (see Fig. 5-4a).

The equations for impact motion are most often written for point mass (zero diameter) particles approaching a cylindrical object. The reason is that most of the experimental work has been done with single wires or fibers and the usual filter medium is fibrous. In dimensionless form for viscous flow, these equations are

$$2\psi \, d^2\tilde{x}/d\tilde{t}^2 + d\tilde{x}/d\tilde{t} - \tilde{v}_x = 0 \qquad\qquad (5-48)$$

and

$$2\psi \, d^2\tilde{y}/d\tilde{t}^2 + d\tilde{y}/d\tilde{t} - \tilde{v}_y = 0,$$

where $\tilde{x}$ and $\tilde{y}$ = transformed coordinates = x/D_c and y/D_c, $\tilde{t}$ = time parameter = $2 \, v_o t/D_c$, v_o = air velocity upstream of obstacle $\tilde{v}_x$ and $\tilde{v}_y$ = velocity transforms = v_x/v_o and v_y/v_o, ψ = impaction parameter (defined below), D_c = diameter of cylindrical obstacle, and t = time. The equations for other regions should have ψ

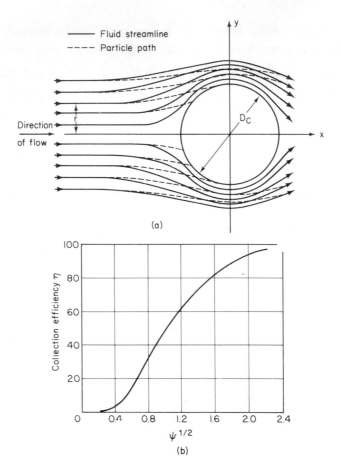

(a)

(b)

FIG. 5-4. Impaction on circular cylinders. (a) Paths of travel. (b) Experimental results $(13 < N_{Re} < 330)$. From J. B. Wong and H. F. Johnstone, Collection of Aerosols by Fiber Mats, Univ. of Illinois Eng. Expt. Sta. Tech. Rept. II(C00-1012), October 1953.

multiplied by the ratio of the Stokes N_{Re} value to the N_{Re} value for the other region in each case. The impaction parameter (ψ) is

$$\psi = C\, \rho_b d^2 v_o / 18 \mu D_c, \qquad (5\text{-}49)$$

where C = empirical correction factor and other terms are as previously defined for Stokes settling. The equations for impaction can be solved by numerical integration. If the limit of collection for a given particle is Y as shown in Fig. 5-4a, the efficiency of collection should be 100% for that type and size of particle approaching within a distance r of the center line of the obstacle, and for any size particle the ratio $2r/D_c$. r depends upon ψ and N_{Re}. Figure 5-4b shows the collection efficiency plotted against ψ to the 1/2 power.

Brownian impaction or diffusive deposition becomes important for collecting small (<1 μm) particles from an airstream that is moving by an obstacle. The important assumptions are the effective collecting zone and the residence time of particles in this zone. Bosanquet assumed that the effective zone is one-half the perimeter of the collecting cylinder. His equation for collection efficiency is

$$\eta_d = \sqrt{(8\mathcal{D}/v_o D_c)} = \sqrt{8/Pe} , \tag{5-50}$$

where η_d = diffusive collection efficiency, $\mathcal{D}$ = diffusivity, and Pe = Peclet number = $vD/\mathcal{D}$ = Reynolds number $(\rho dv/\mu)$ times the Schmidt number $(\mu/\rho\mathcal{D})$.

Interception is the term given to the collection of particles on fibers as a result of the size of the particles, as opposed to the point mass of the impaction. Ranz and Wong assumed potential flow and obtained the equation for interception efficiency of cylinders as

$$\eta_n = 1 + R - 1/(1 + R) , \tag{5-51}$$

where η_n = efficiency of interception and R = diameter of particle divided by diameter of cylinder = d/D_c. For spheres as collectors, the $1 + R$ term in the numerator of the last term of Eq. 5-51 should be squared.

Davies has combined the effects of impaction and interception and their dependence upon one another and developed the following equation.

$$\eta_{dn} = 0.16\,[R + (0.50 + 0.8R)\psi - 0.1052\,R\psi^2] \qquad (5\text{-}52)$$

7. Thermal Migration

Particles are moved from the hot toward the cold in a thermal gradient. Supposedly, two forces are responsible--a more energetic bombardment by the gas molecules on the hot side than on the cold side and a pinch effect from the migration of gas molecules on the surface from the cold toward the hot. Various formulas for the thermal movement of particles have been put forth, and of these probably the most successful is one by Epstein for heat-conducting spheres;

$$F_t = -9\pi d/2\,[k_g/(2k_g + k_p)]\,(\mu^2/\rho_g T_g)\;dT/dx, \qquad (5\text{-}53)$$

where F_t = thermal force on particle (dynes), d = particle diameter (cm), k_g and k_p = thermal conductivities of gas and particles (cal/sec-cm-$^\circ$K), ρ_g = density of gas (gm/cm^3), T_g = absolute temperature of gas ($^\circ$K), μ = dynamic viscosity of gas (poises), and dT/dx = thermal gradient ($^\circ$C/cm).

8. Electrical Migration

Charged particles are moved in an electrical field by attractive and repulsive forces. Moreover, a particle that enters an electric field becomes charged. These facts are used to collect particles from airstreams for sampling and air cleaning. The migration of a charged particle in an electric field is calculated from

$$\Omega = E_o E_c d/4\pi\mu = 0.884 \times 10^{-4}\,E_o E_c d/\mu, \qquad (5\text{-}54)$$

where Ω = migration velocity (cm/sec), E_o and E_c = intensity of charging and collecting fields--often same (esu in left equation and kV/cm in right), d = particle diameter (cm and μm), and μ = dynamic viscosity of air (poises). Equation 5-54 has several assumptions incorporated in its derivation; however, it serves a practical purpose in electrostatic precipitation. Two multiplying corrections often needed are one for the dielectric constant (ξ) which is $3\xi/(\xi + 2)$, and Cunningham's correction (Eq. 5-43) for particles less than a few micrometers in diameter.

9. Other Migration Forces

In addition to the migration forces already described, there are some lesser forces that cause movement of particles in an aerosol, including photophoresis and diffusiophoresis.

Photophoresis is the movement of particles by the unequal bombardment of photons when the particle is lighted from one side. The actual force is caused by the rebound of the gas molecules from the hotter (illuminated) side of the particle with greater velocities.

Diffusiophoresis is the movement of particles by a flux of water vapor moving toward a condensing surface. The particles are moved along by the interaction with the flux. Such movement accounts for a clear space surrounding a water droplet during condensation. This movement is hindered by Stephan flow which compensates to some extent for the flux movement.

III. SUMMARY

Knowledge of the basic properties of gaseous and airborne particulate matter is often requisite to the performance of a good job in air sampling and analysis. Most of the material mentioned here can be found more fully described in a physical chemistry or other books listed in the bibliography.

The collection and analysis of material may involve a change in state--melting, freezing, sublimation, absorption, adsorption, or condensation. In addition, problem solving in air pollution is often based upon transport properties, heat content and conduction, laws of work, and aerosol mechanics. All of these facets are reviewed in a manner that may jog the memory but do not provide adequate first coverage.

PROBLEMS

1. What are the limits, if any, on the applicable range of the ideal gas relation $P_1V_1/T_1 = P_2V_2/T_2$?

2. What are the possible positions of the three sampling train components relative to each other when the items are: (1) a vacuum pump, (2) a filter with a 700-mm mercury head drop, and (3) a critical orifice to control the flow? Why is a meter needed instead of using the filter itself as a critical orifice since 700 mm mercury is greater than 0.528 atm?

3. Why is fiber glass insulation more effective than leaving the container in contact with the air when in both instances air is the insulating material and there is more air present without the fiber glass than with it?

4. How much activated carbon is needed to collect several micrograms of phenol? Should activated carbon be used for collecting ethylene? Why?

5. What is the theoretical migration velocity for a 10-μm ferric oxide particle under 600 g's?

6. What is the migration velocity for a 1-μm limestone particle in an electrostatic precipitator with a charging and collecting potential of 16 kV/in.?

7. What is the head loss (in. w.g.) in 70 equiv. ft of rough duct which is 18 in. diameter and at 6 in. w.g. pressure with 7000 scfm flow of air?

BIBLIOGRAPHY

C. H. Bosanquet, Appendix to "Dust Collection by Impingement and Diffusion," (C.J. Stairmand), Trans. Inst. Chem. Eng. (London), 28, 130-139, 1950.

R. D. Cadle, Particle Size, Van Nostrand Reinhold, New York, 1965.

C. N. Davies, "The Separation of Airborne Dust and Particles," Proc. Inst. Mech. Eng. (London), 1B, 185-198, 1952.

P. Drinker and T. Hatch, Industrial Dust, 2nd ed., McGraw-Hill, New York, 1954.

C. W. Eshbach, Ed., Handbook of Engineering Fundamentals, 2nd ed., Wiley, New York, 1952.

B. R. Fish, The Electrostatic Forces of Adhesion between Particles, ORNL-TM-1947, Oak Ridge National Laboratory, Oak Ridge, Tennessee, August 1967.

N. A. Fuchs, The Mechanics of Aerosols (transl. by R. E. Daisley and Marina Fuchs; C. N. Davies, Ed.), Pergamon Press, New York, 1964.

G. Herdan, Small Particle Statistics, 2nd rev. ed., Butterworths, London, 1960.

C. D. Hodgman et al., Eds., Handbook of Chemistry and Physics, 50th ed., Chemical Rubber Publ. Co., Cleveland, Ohio, 1968.

E. Hutchinson, Physical Chemistry, Saunders, Philadelphia, Pennsylvania, 1962.

I. Langmuir, "Supercooled Water Droplets in Rising Currents of Cold Saturated Air," The Collected Works of Irving Langmuir, 10, 199, 1944.

P. A. Leighton, Photochemistry of Air Pollution, Academic Press, New York, 1961.

H. Neuberger, H. Panofsky, and Z. Sekera, "Physics of the Atmosphere," Air Pollution Handbook, P. Magill, F. Holden, and C. Ackley, Eds., Section 4, McGraw-Hill, New York, 1956.

W. E. Ranz and J. B. Wong, "Impaction of Dust and Smoke Particles on Surface and Body Collectors," Ind. Eng. Chem., 44:6, 1371-1381, June 1952.

A. Schütz, "The Electrical Charging of Aerosols," Staub (English), 27:12, 24-32, December 1967.

W. Strauss, Industrial Gas Cleaning, Pergamon Press, New York, 1966.

A. Turk et al., "Determination of Gaseous Air Pollution by Carbon Adsorption," Am. Industrial Hyg. Assoc. Quart., 13, 23, 1952.

H. C. van de Hulst, Light Scattering by Small Particles, Wiley, New York, 1957.

Chapter 6

SAMPLING--THEORY, EQUIPMENT, AND TECHNIQUES

I. INTRODUCTION

A good job of air sampling requires careful planning, equipment preparation, and procedures that adequately serve the purpose. Often the greatest handicap to be overcome is insufficient funds that have been allocated according to a gross underestimate of the real difficulty involved in sampling and analysis. The time, effort, and know-how necessary to insure successful measurements are usually well-rewarded in applying the measurements to control design.

II. PLANNING FOR AIR SAMPLING

It is worthwhile to expend a fair amount of effort in planning for air sampling. Effective planning is based upon knowledge of the operation--the purpose of the sampling, what is to be measured, its source, order of magnitude, concentration and degree of mixing, the analytical method to be used, interferences in either the sampling or the analysis, and the statistical development of the sampling parameters. The factors to be decided are the location, time, rate, amount, sampling method, and efficiency of the sampling.

A. Preplanning Considerations

The purpose of the sampling often dictates the conditions of collection. The purpose of most ambient air sampling is to test for compliance with regulations, to monitor the exposure being received by people, animals, or plants, to measure the dispersion and/or decay of pollutant in the atmosphere, or to

195

determine concentrations for research or training purposes.
Stack or duct sampling is made to check emissions for compliance
with regulations, for efficiency of stack gas cleaners, or for
process control.

Regulations frequently specify not only the maximum allowable concentration or emission but also the method of measurement, sampling and analysis, for checking compliance. Once a
method has been standardized for a given purpose, it is difficult
to change the method even when better methods become available;
for example, the number of dust particles per cubic foot for silicosis risk was for a long time sampled only with the midget impinger despite its inefficiency in collecting small particles;
this was done because the incidence of silicosis was correlated
with concentrations measured by this means.

The consideration of purpose in sampling has led to the
development of many special-purpose samplers. For example,
the respirable dust sampler has a cyclone which presumably
collects particles that would be removed in the upper respiratory
tract, followed by a filter which removes particles that would
reach the lower respiratory system.

B. Sampling Location

The location of the sampling point is often quite important
to the results obtained. Careful thought should be given to the
subject before deciding just where to sample. Places where the
pollutant may be concentrated by stilling action or special currents should be avoided as much as possible. Sampling should
not be done in the downwash of buildings or topographical features. Neither should obstructions limit the amount of pollutant
reaching the sampler. A rule of thumb which has been used for
siting rain gages may be applied for cramped quarters--a 45°
elevation angle swung about the sampler should describe a cone
about the vertical that is free of obstructions; some observers
say use a 60° angle.

In most cases even more important than obstructions or concentrating factors is the sampling height. Sampling airborne particulates at a level that picks up resuspended particles shows much higher readings than those made on rooftops or sampling towers at a distance well above the ground. Direction of sampler orientation is important (discussion in Section IV, C). The height of sampling should be selected to satisfy the philosophy of the sampling. If exposure to people is the desired measure-ment, sampling should be carried out at the height of an average person. Presumably, this includes consideration of whether the material will be inhaled or ingested for internal exposures to toxic materials, or received throughout the body for external exposures to penetrating radiations. The ultimate in sampling location may be the use of a personal sampler with the probe positioned adjacent to the nose for sampling inhaled pollutants. Ambient sampling is usually carried out at a height of 15 to 50 feet.

C. Sampling Time

For effective results the time of sampling and the sampling time (when to sample and how long to sample) must be chosen carefully with due regard to the purpose of sampling, the secu-lar variations of the source emissions, and the degree of mixing the source receives before reaching the sampling location.

If the purpose of sampling is to determine the peak concen-tration of pollutant, sampling time must be quite short, but repeated continually over a sufficient interval to insure finding the peak concentration. For example, the carbon monoxide con-centration in a city reaches peaks during the morning and evening rush hours on week days; therefore peak sampling must be carried out at those times, but sampling for average concentration re-quires at least 1 week of sampling or 1 weekday plus a weekend, with the assumption that the weekday is typical. The cycling

interval and period of persistence for source emissions vary
from short-interval instantaneous releases, such as boiler blow-
down, to sources occurring with yearly cycles and persisting
for months, such as space heating. In order to define the concentration peak, a short sampling interval is repeated continually
over the time known to contain the peak. The sampling time must
be sufficient to collect the amount of pollutant needed for analysis.

As a result of incomplete mixing during transport (see Chapter 3, Section V, E), the peak concentration found depends upon
the sampling time, the shorter sampling time giving the higher
peak. The magnitude of the concentration fluctuations is determined by the degree of mixing, that is, buoyant stability,
mixing time or distance from the source, source geometry, and
roughness of the topography. Those conditions that promote
rapid dispersion cause large variations of concentration close
to the point of emission. Most investigators who have taken
air samples have noticed the time dependency of the peak concentration. Among those who have reported the magnitude of
these fluctuations, there is wide disagreement, which probably
results from the different conditions during the samplings. The
only strict agreement seems to be that a power relation exists.
Singer and co-workers used a logical approach to describe why
there should be a power relation between the peak concentration
found and the sampling time used. On the basis of their evidence
and that of nine other reports, reasonable estimates of the peak-
to-mean concentration ratios (C/C_0) for various sampling time
ratios (t/t_0) seem to be as shown in Fig. 6-1. A frequently used
time base (t_0) has been 1 hr, but times from 1 min to 24 hr have
also been used. Quite obviously, the curves cannot extend indefinitely with time at the slopes shown, but must flatten out
for very long times (more than a few days) and very short times
(less than a few seconds). Larsen has recently developed a model
from the log normal to extend the curves; perhaps extremal

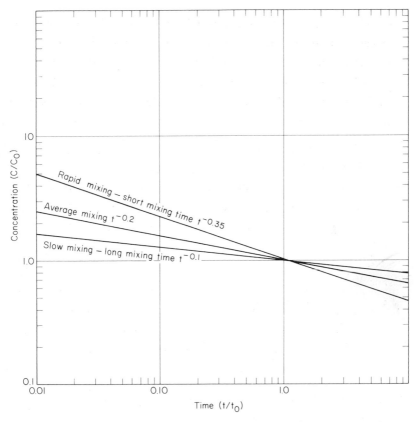

FIG. 6-1. Effect of sampling time on peak concentration.

statistics would better serve the purpose. Fumigation gives the steepest slope, maybe near the vertical limit; and laminar flow over a smooth surface, very small eddies for a long time, or a very large (areal) source with uniform emissions gives the flattest slope to the variation plot, the limit being horizontal.

EXAMPLE 6-1: A concentration of 0.70 ppm of sulfur dioxide is measured for a 10-min sample during a time of neutral stability. Would this concentration probably comply with a regulation of 0.5 ppm_v for 30 min?

$t/t_0 = 10/30 = 0.33$ from Fig. 6-1, $C/C_0 = 1.25$

$C_0 = 0.7/1.25 = 0.56$, which means probable violation.

The minimum concentrations found also vary with the
sampling time, the lower minimum (C/C_0) occurring with the
shorter sampling time. Larsen found the oxides of nitrogen in
Washington, D. C., to follow a power of -0.29 for peak concentrations and +0.36 for minimum concentrations (valleys).

D. Size of Sample

The size of the sample required depends upon the concentration of the material being sampled and the analytical method
to be used for the analysis. The size may range from considerably less than a liter to many cubic meters and may contain from
a picogram (10^{-12} gm) or less to several grams of the material
for analysis.

The determination of how much sample to collect depends
upon a knowledge not only of the analytical sensitivity to be
used but also the approximate concentration of the component
of interest. The relative sensitivities of analytical methods are
shown in Table 7-1. If no more definitive information is available
the normal concentrations found in the urban air by the NASN may
be used for an estimate of the expected concentration (see Table
1-3). When this information is not suitable, exploratory sampling
may be carried out for a qualitative and roughly quantitative
answer which can be used for planning better sampling determinations.

The size of the sample required can be statistically determined from estimates of the pollutant concentration and the
standard deviation that obtains in the measurement of that component. If no better estimate of the standard deviation is available from past experience in the locale, assume that it is 25%
of the concentration being measured. The approach to be used
relates the sample standard deviation to the population standard
deviation. For ambient air sampling the population is assumed

to be infinite and the estimate of the population standard deviation is made by

$$s \approx \sigma /\!/N, \tag{6-1}$$

where s = the population standard deviation, σ = the sample standard deviation, and N = the size of the sample.

EXAMPLE 6-2: What size sample is required for a 90% confidence limit so that the error in urban sulfate concentration will not exceed 10%?

$ks = |\bar{x} - \mu| \leq 10\% \, \mu$ k = 1.64 for 90% confidence

$\sigma = 0.25$ (estimated) $\mu = 10.6 \, \mu g/m^3$ (from Table 1-3)

$1.64 \, (0.25 \times 10.6)/\!/N \leq 0.10 \, (10.6)$

$/\!N \geq 4.1$ $N \geq 16.8 \, m^3$

For a finite population such as the air inside a room, the same type of approach is made by employing the binomial distribution.

EXAMPLE 6-3: How many 1-gm samples should be taken out of a 1-kg dust sample to obtain a standard deviation fraction of 10% if the attribute tested is 1% of the total weight?

$$n = Npq/[\sigma_f^2 (N - 1) + pq]$$

where n = required sample size, N = size of population, σ_f = standard deviation as fraction of n, p = fraction of sample with attribute being tested, and q = 1 - p.

$$n = 1000 \, (0.05)(0.95)/[0.10^2 \, (999) + 0.05 \, (0.95) = 4.7$$
or 5 samples

In some cases simple reasoning can indicate the proper sample size. For example, weights made on analytical balances may be incorrect by 3 units in the last digit. If it is desired to keep this error less than 10%, 30 units of particulates should be collected. For the usual balance this weighing error is in the 10^{-4}-gm position and 3 mg is the minimum sample needed.

E. Sampling Rate

The sampling rate must be selected to fulfill the requirements of sample size and sampling time. Furthermore, the sampling rate must satisfy the velocity or flow rate through the collector that fits the calibrated conditions. Table 6-1 lists the normal sampling rates for a number of samplers.

A sampling rate may be determined by the amount of sampled material required for analysis and the concentration of such material in the air. The rate is selected to give sufficient material within a reasonable sampling time. This is especially true for gravimetric determinations of low concentrations that indicate high-volume sampling.

F. Collection Efficiency

In very few instances is 100% sampling efficiency desirable. Usually total collection is made at a sacrifice in flow rate

TABLE 6-1

Sampling Rates for Various Samplers

Sampler/collector	Usual rates	
	liters/min	cfm
High-volume sampler	--	45-70
Electrostatic precipitator	--	0.1-5
Whatman No. 41 2-in. filter	10-120	--
Andersen sampler	--	1
Greenburg-Smith impinger	--	1
Cascade impactor		
Unico	10-40	--
Casella	17.8	--
Membrane filter (2-in. diameter)		
5-μm Pore size	40-50	--
0.3-μm Pore size	3-5	--
Paper tape sampler	5-30	--
Midget impinger	--	0.1
Thermal precipitator	0.02-0.1	--

or at a higher cost for equipment or with a compromise in the
maintenance of the condition of the sample. For instance,
according to efficiency curves on membrane filter collection,
a 5-μm pore size collects > 98% of 0.3-μm particles at a flow
rate more than 10 times that of a 0.3-μm pore size which collects 100% of the particles that size. As a result, sampling
time could conceivably be reduced from 2 hr to 15 min by using
the larger pore size.

It is usually desirable to know the collection efficiency
so that absolute amounts can be determined. This is nearly
always true for sampling in connection with research; however,
sometimes qualitative information gives all the definition of a
pollutant needed. A notable exception to the need for knowledge
of sampling efficiency is a standard test, one to which effects
have been related and by which standards have been set, such
as the midget impinger for sampling silica dusts.

III. AIR SAMPLING EQUIPMENT

Most air sampling involves passing air through a device
that collects the material of interest from the air. Other sampling may merely collect a portion of the air mixture or pass the
stream through analysis on the spot. Whichever sampling method
is employed, an air mover is usually needed, exceptions being
dustfall, sulfation rates, corrosion, and other such studies that
passively sample the air moved by wind and circulation. The
air mover should be well-suited to the process in terms of flow
rate at the head loss encountered. In normal circumstances,
the air flow through the collector is induced; that is, the collector is on the vacuum side of the air mover. This avoids the
possible addition or removal of materials in passing through the
pump.

A. Air Movers

Probably the most widely used air mover in sampling is the carbon vane pump. This pump has a rotor which is set off-center in a cylindrical cavity; the rotor usually has four carbon vanes that are kept against the outer walls of the chamber by centrifugal force and/or spring loads; the vanes divide the chamber into compartments which change in volume as the rotor turns. The change in volume of the compartment results in a change in pressure and moves the air. Carbon vane pumps can be run continuously for hours or days. A squirrel cage motor is normally used to drive these pumps from 1800 to 6000 rpm. The pumps pull 18-20 in. of mercury when the vanes are not worn and move up to about 2 cfm of air. Larger sizes of vane pumps are available. Although the volume moved is nearly constant (positive displacement pumps), their corrected volume moved (scfm) does vary with the head (pressure X volume = constant). Figure 6-2a shows a characteristic curve. The pumps are frequently equipped with needle valves to admit external air and adjust the flow through the sampling line. These pumps cost about $100 and at least one is practically necessary to any air sampling activity. Their versatility is enhanced by the fact that they can be used as pressure pumps up to about 15 psig.

Various piston pumps are applied to air sampling. The usual ones are motor-driven and consist of a piston or multiple pistons connected to a crankshaft or a cam-type race ring. The motors are normally continuous duty and the pump can operate for long periods without being attended. Although larger sizes are available, the usual piston pump for a laboratory has a maximum of 2-4 cfm capacity and a maximum operating head of 27 in. of mercury. The cost is about $100 but may be increased by accessories such as pressure or vacuum gages and a rotameter.

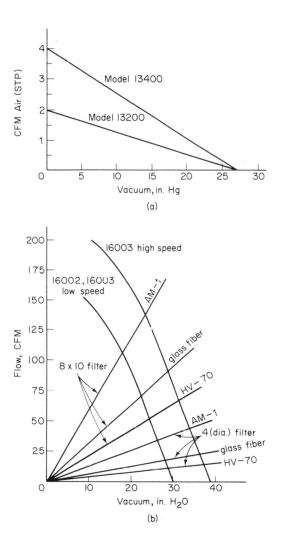

FIG. 6-2. Air sampling pump characteristics. (a) Carbon vane. (b) Centrifugal. (Gelman oil-less vacuum pumps and Hurricane Samplers).

Various hand-operated pistons such as hypodermic syringes and pumps for pulling small volumes through indicating glass tubes (see Chapter 7) are quite practical for moving small volumes of material and in accurately measured amounts. Also, rubber bulbs and various bellows are used for moving small volumes of air against low or moderate heads.

Other pump configurations are sometimes used for positive displacement of air samples. These include lobe pumps, screw pumps, diaphragms, and roller pumps (squeezing flexible tube chambers). Lobe pumps are similar to carbon vane pumps in application; the last-mentioned three are for small flow rates and moderate heads. Mylar bags are filled to a definite volume by holding them in a container evacuated by some type of pump. The rate of fill of the bag is about the same as the pump rate, especially after start-up. Recently, spring-loaded Mylar bags have been devised that fill themselves.

Centrifugal pumps are used for moving large volumes of air against moderate heads. The original high-volume pumps were simply the mechanisms from vacuum cleaners, but actual sampler designs are now being built. A high-volume sampler normally has a maximum capacity of up to 200 cfm and moves up to about 70 cfm (see Fig 6-2b) at 30-40 in. w.g. (water gage). The high-volume samplers on the market range in price from about $100 to $200. One big drawback of this type of sampler, and all others with brush motors, is that maintenance is often necessary to keep the motor operating.

Moderate to large volumes can be moved against low heads with axial flow fans--the normal propeller-type fan. Such a fan is applicable to electrostatic precipitator collectors with their flows of a few cubic feet per minute and head losses of approximately 0.1 in. w.g.

An ejector pump is practical for some applications, especially those in which a nonsparking power source is required or those in which tap water or compressed air is more readily available than electrical outlets. The Uni-Jet (Unico) sampler (about $150) powers a midget impinger for 90 min on a 1-lb can of Freon ($1.75). The flow rate is regulated to be practically constant over the entire sampling period.

Battery-operated pumps are available. Wet batteries, such as auto batteries, have been a popular power source but these are being replaced by nickel-cadmium rechargeable dry batteries. The midget or lapel sampler has a diaphragm pump which moves up to 3.5 liters/min through a paper filter or an impinger. The pump, battery, and flowmeter are in a pocket-sized case and weigh less than 1 lb. The battery powers the unit for about 10 hr on one charge.

One of the most inexpensive pumps available for moving air is the diaphragm pump used to aerate aquariums. Pet stores often furnish these pumps at a fraction of the cost of similar pumps from other sources.

Mine Safety Appliances (MSA) sells a four-cylinder, portable, hand-cranked pump for use with the midget impinger. The pump has a vacuum gage that reads inches of water. Cranking at about 50 rpm produces a vacuum of about 12 in. w.g. and gives a flow of 0.1 cfm through the midget impinger.

Any vacuum source may be used as an air mover for sampling purposes--the intake manifold of an auto engine, an aspirator on a water spigot, an evacuated flask, or water (other liquid) displacement.

Automatic flow rate adjusters have been built to keep a constant flow rate through a high-volume sampler. This is not

considered critical, and averaging the beginning and ending
flows does not introduce serious errors.

B. Meters

The determination of the concentration of a collected pollutant requires that the volume that contained the sampled material be known; that is, the volume must be metered. The three
types of measurement used are volume in an elapsed time, volumetric flow rate, and velocity in an enclosed area. Measurements of the first type are made with spirometers, wet test and
dry test meters, rubber bulbs, evacuated containers, and piston
pumps and syringes. Volumetric flow rate is determined by
using rotameters, orifices, and other devices that measure the
velocity in a known small section. The flow rates (velocities)
in a large section are determined with pitot tubes, deflecting-
vane or hot-wire anemometers, or other similar instruments.

1. Spirometer

A spirometer is a very accurate volume-measuring device
and is therefore used as a primary standard for calibration purposes. A spirometer employs direct positive displacement of a
calibrated cylinder which moves with a water seal (see Fig. 6-
3a). The most elaborate spirometers give a constant rate of flow
as controlled by counterweights; the effect of the buoyancy on
the cylinder wall is even removed by a small weight with a
variable-moment arm.

2. Wet Test Meter

The wet test meter uses water displacement to rotate a
segmented drum which is geared to indicating dials to count
the revolutions of the drum (see Fig. 6-3b). The wet test meter
can give volume measurements with an error of about $0.1-0.5\%_v$,
even for very small flow rates. The usual maximum flow rates

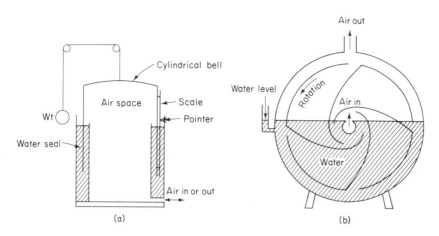

FIG. 6-3. Laboratory flow meters. (a) Spirometer. (b)
Wet test meter.

are from a few tenths to more than 1 cfm. The wet test meter
is not suitable for field use but is often employed as a secondary
standard for the laboratory, especially for laboratories that do
not have good spirometers.

3. Dry Test Meter

The expansion of a bellows or diaphragm is used to drive
indicator dials for volume measurements. Two sets of bellows
are valved and connected such that one empties as the other
fills. The dry test meter ordinarily gives accuracy within 1 or
$2\%_v$ at flow rates of a few hundredths to 1 cfm. This meter is
not really convenient enough for field usage except in unusual
circumstances. The smallest dial reading division is normally
0.001 ft^3.

4. Rotameters

A rotameter has a float which is driven upward in a tapered
tube (see Fig 6-4a) until the drag force balances the weight of

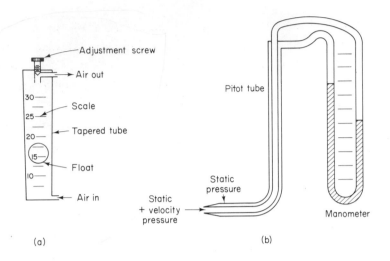

(a)   (b)

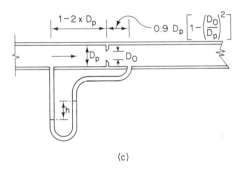

(c)

FIG. 6-4. Field flow meters. (a) Rotameter. (b) Pitot tube. (c) Sharp-edged orifice.

the float--ball, cone, or other shape (see Chapter 7 for drag-equal-weight calculations). The area of the air flow by the float changes to give a constant velocity for the drag. The accuracy of the rotameter is normally within 5-10%$_v$, according to the linearity of the tapered walls, the degree of taper, and the calibration.

The rotameter is the meter of choice for many if not most field measurements. It is rugged and economical (average price about $25). Rotameters are often built into sampling pumps and if kept clean will last for a considerable amount of operation. Their calibration should be checked at frequent intervals. Rotameters are readily available in a wide range of sizes from cubic centimeters per minute to cubic feet per minute.

The principal difficulties with rotameters are oscillation of the float and dirt encrustation after long-term usage or use in dirty gas streams. The oscillation problem occurs primarily in rotameter tubes with long gradual tapers, those intended to measure accurately. Oscillations can frequently be prevented by starting the flow gradually or damped by using a magnet or a constriction in the line. The rotameter can usually be disassembled and washed as required.

5. Pitot Tube

The pitot tube is designed to change the velocity head to a pressure head which can be measured with a manometer. The stream velocity is reduced to zero at the tip of the tube and the resulting pressure change measured with an arrangement such as that shown in Fig. 6-4b). The velocity is calculated from the fluid mechanics equivalence that $v^2 = 2gh$.

EXAMPLE 6-4 : Relate velocity (v) in feet per minute to head (h) in inches of water gage.

$v = \sqrt{(2gh)}$ for units of fps $= \sqrt{[ft/sec^2 \times ft \text{ (of air)}]}$

$v = 60 \text{ sec/min} \sqrt{[2(32.2) h/12 \text{ in. } (62.4 \text{ lb/ft}^3 \text{ water}/\rho_a)]}$

$$v = 1097 \sqrt{(h/\rho_a)}, \qquad\qquad (6-2)$$

where v = fluid (air) velocity (fpm), h = head (in. w.g.), and ρ_a = density of fluid (air) (lb/ft^3). For air at NTP, ρ_a is 0.75 lb/ft^3 (from Table 1-2) and the relation becomes

$$v = 4002 \sqrt{h}.$$

6. Orifices

As noted for the pitot tube, a change in velocity of a fluid is accompanied by a change in pressure. The orifice constricts the flow to change the velocity, and the resulting pressure change or head loss is used to determine the flow rate. The orifices commonly used are the sharp-edged (subcritical) orifice and the critical orifice.

The sharp-edged orifice can be analyzed theoretically, but it is necessary to calibrate the orifice at various flow rates because of the head loss that is not recoverable. The equation for the flow rate through the orifice is

$$Q = 1097 \ CA\sqrt{(h/\rho)}, \tag{6-3}$$

where Q = flow rate (cfm), C = coefficient of head loss ≈ 0.60 for usual conditions, and A = area of orifice (ft^2). The locations for the manometer connections across the orifice are made in consideration of the flow regimes, and the recommended locations are shown in Fig 6-4c. Various orifice configurations are available. One make of high-volume sampler uses a set of interchangeable orifices and a Bourdon tube pressure gage on the exhaust from the sampler. For relatively small flow rates, a glass manometer with four different sizes of orifices in a rotatable glass turret is on the market.

An orifice that has the downstream pressure loss less than 0.53 of the upstream pressure operates critically. The velocity through a critical orifice is sonic and varies little even for large pressure changes.

EXAMPLE 6-5: Derive the pressure conditions for the critical orifice.

From fluid mechanics, sonic velocity (c) in a compressible medium is

$$c = \sqrt{(g\gamma pV)},$$

where g = acceleration of gravity, γ = ratio of specific heats = c_p/c_v, p = pressure, and V = specific volume of gas. For isentropic gas expansion

$$(v_2{}^2 - v_1{}^2)/2g = J c_p (T_1 - T_2),$$

where v = velocity (subscript 2 at orifice and 1 upstream), J = entropy of gas, T = absolute temperatures, c_p = specific heat at constant pressure, and g = acceleration of gravity.

Substitute $c_p = \gamma R/[J(\gamma - 1)]$ and $pV = RT$ to obtain

$$(v_2{}^2 - v_1{}^2)/2g = [\gamma/(\gamma - 1)] (p_1 V_1 - p_2 V_2).$$

Neglecting v_1 and substituting c equivalence for v_2 gives

$$g p_2 V_2/2g = [\gamma/(\gamma - 1)] (p_1 V_1 - p_2 V_2).$$

From $pV^\gamma = k$ for adiabatic expansion,

$$V_1/V_2 = (p_2/p_1)^{1/\gamma}.$$

Then
$$p_2 = [2/(\gamma + 1)]^{\gamma/(\gamma - 1)} p_1.$$

$\gamma = 1.4$ for air (diatomic molecules) at NTP and

$$p_2 = (2/2.4)^{1.4/(1.4 - 1)} \qquad p_2 = (0.833)^{3.5} p_1$$

$$p_2 = 0.528 p_1$$

The area of the critical orifice is selected to give the desired flow rate (sonic velocity ≈ 1140 fps at NTP). Because of the difficulty of drilling very small holes, hypodermic needles (which are available in many sizes) have been useful in obtaining small flow rates by critical orifices.

7. Deflecting-Vane Anemometer

The deflecting vane anemometer makes use of the fact that wind exerts a pressure on a surface. The variation of surface pressure with the angle of incidence of the airstream and probes that admit varying fractions of the flow through to the vane permit measurements of a rather wide range of velocities with a single instrument. The weight of the vane or the stiffness of a spring load can also be varied to obtain a wide-range capability.

8. Other Metering Methods and Devices

Other metering methods and devices are possible and are
sometimes used. The deflection of a beam of particles traveling
from a ^{210}Po source to a detector is proportional to the velocity
of the air traversed. The drag force on a tethered object, such
as a string, can be measured. The dilution of a tracer substance
(radioactive or stable) injected at a known rate into an airstream
gives a measure of flow rate. The dilution rate of a tank of pure
gas can be used to measure airflow (see Section IV, A). A propeller anemometer can be used to measure velocity of flow. The
rates of sound transmission with and against air movement can
measure velocity. The velocity of a soap bubble in a buret can
be used to measure very low flow rates. Venturi meters are
sometimes used for high-volume flow measurements.

Pumps sometimes also serve as metering devices. For
example, piston pumps and rubber bulbs give a definite volume
for each stroke or expansion. Hypodermic syringes (piston
pumps) in particular are often used for measuring gas volumes
into dilution chambers for odor or other tests.

The hot-wire anemometer (see p. 93) can be used to measure air velocities across a duct for determination of volumetric
flow rates. The hot-wire anemometer should not be used in
dirty gas streams, however.

C. Sample Collectors

All of the principles used in air cleaning for removal (not
destruction) of particulate and gaseous pollutants are used for
sample collection. In addition, some methods that are impracti-
cal for air cleaning are practical for air sampling, for example,
thermal precipitation. Much higher head losses can be tolerated
in a sampler than in a cleaner where power costs would be prohibitive for large flows and long periods.

1. Filters

Most particulate sampling is made with some kind of filter. In addition, filters may be impregnated with chemicals that react with certain gases and are used to absorb gases as well. The filter materials in common usage for air sampling are membrane, paper (cellulose fiber), fiber glass, felt, plastic fiber, and mixtures of fibers. More exotic filters used for special purpose research are filters with cylindrical holes (Nuclepore) and membrane filters made of a silver matrix.

Filters remove particles by three primary means--impaction, sieving, and diffusion or Brownian motion. Some types of filters develop high electrostatic charge from the movement of air over the material; the charge adds greatly to the efficiency of membrane, fiber glass, and some plastic materials. Diffusion is an effective means of removal only when the particle diameter is below a few tenths of a micrometer and then only if the residence time of the particle in the filter matrix is considerable, that is, low velocities and thick filters. The relatively minor role of straining as compared with impaction can be illustrated by the fact that Whatman No. 41 filter paper with an effective pore size of about 20 μm removes 98% of particles about 0.2 μm in diameter provided the face velocity is about 200 fpm (100 cm/sec). A similar comparison with electrostatic forces is found in a 5-μm-void-size glass fiber filter that collects 99.992% of DOP (dioctyl phthalate, 0.3 um diameter particles) smoke regardless of collecting face velocity (within limits). If filtration is continued for a very long period, there will be some migration of the particles collected by impaction and electrostatic forces, but this is of no real concern in air sampling.

There is disagreement as to best filtering (face) velocities; however, most of the tests that have been made on impaction-type filters (Whatman paper) point to a face velocity of about

TABLE 6-2

Efficiencies of Some Popular Filters[a]

Filter	Efficiency on 0.3-μm DOP particles			Efficiency on 0.025-μm MMD particles[b]		
	Maximum %	Flow (fpm)	Head (in. w.g.)	Maximum %	Flow (fpm)	Head (in. w.g.)
Whatman No. 41	85-98.6	200	26.5	75-84	5-200	to 26.5
Glass fiber 1106B	99.5-99.9+	5-300	0.7-48	99-99.9+	5-200	to 32
HV-70 (9 mil)	96.5-99.7	5-300	0.7-56	95.2-99.1	5-200	to 39
Millipore AA (0.8-μm pore)	98.5-99.9+	5-300	to 225	99.9+	5-200	to 150
Gelman AM-4 (0.8-μm pore)	98.6-99.9+[c]	5-200	to 160	99.7-99.9+	5-200	to 160

[a]Adapted from various sources in bibliography.   [b]Uranine particles.

[c]Uranine particles with MMD = 0.27 μm and σ_g = 1.8.

200 fpm (100 cm/sec) as giving maximum efficiency. Table 6-2
shows efficiencies for some popular filters at given flow rates
and head losses. Table 6-3 lists pertinent information about
filter characteristics, including head losses. For other flow
rates, head losses can be plotted as straight lines through the
origin (0,0) when flow is plotted against head loss either on
arithmetic graph paper or log log paper.

   a. Membrane Filters. Membrane filters may be thought
to consist of a maze of branched fibers between two flat surfaces
which have numerous pores (see Fig. 6-5). Most of the collection takes place at the surface, a factor that is important in
sampling particles for gross visual analysis or for counting short
range radiations from collected particles. The specified pore
size is the maximum pore size and is available over a wide range
(0.004-10 um or more). Quality control on the manufacture of
membrane filters is quite good and one can rely on the specified
pore size not being exceeded. The 0.8-um pore size is probably
the most used size for air sampling; this filter achieves high
efficiencies even on small particles because of the tortuous path
of the air through the filter.

   Membrane filters can be made from any of several plastic
materials or even from silver, but they are commonly esters of
cellulose, especially cellulose acetate. Special filters are
available that withstand heat, acid, hydrocarbon solvents, or
other chemicals. Filters to be used for gravimetric analyses
must be suited to that purpose because the evaporation of solvent
remaining in the usual filter actually causes a weight loss as air
is passed through the filter. By placing two filters in the holder
and assuming that the weight loss shown by the back-up filter
occurred in the front filter as well and that the front filter collected 100% of the particles, it is possible to do a fair job of
gravimetry with the common filters.

TABLE 6-3

Filter Characteristics[a,b]

Filter designation	Void size (μm)	Fiber diameter (μm)	Thickness (μm)	Weight per unit area (mg/cm²)	Ash content (%)	Temperature ceiling (°C)	Tensile strength	Head loss at 100 fpm (in. w.g.)	Source[c]
				Cellulose fiber filter					
Whatman 1	2+	NA	130	8.7	0.06	150	NA	40.5	RA
4	4+	NA	180	9.2	0.06	150	NA	11.5	RA
32	1–	NA	150	10.0	0.025	150	NA	38(28 fpm)	RA
40	2	NA	150	9.5	0.01	150	NA	54	RA
41	4+	NA	180	9.1	0.01	150	NA	8.1	RA
42	>1	NA	180	10.0	0.01	150	NA	46(28 fpm)	RA
44	>1	NA	150	8.0	0.01	150	NA	40(28 fpm)	RA
50	1	NA	100	10.0	0.025	150	NA	49(28 fpm)	RA
541	4+	NA	130	8.2	0.008	150	NA	NA	RA
S and S 604	NA	NA	200	NA	0.03	80	NA	8.5	S and S
MSA Type S	NA	NA	1000	NA	NA	120	NA	NA	MSA
Cellulose	NA	NA	1000	NA	NA	120	NA	NA	MSA
Corrugated cellulose	NA	NA	1000	NA	NA	120	NA	NA	MSA
IPC 1478	NA	17 avg.	760	14.6	0.04	120	Woven backing	0.31	IPC

Glass fiber filters

MSA 1106B[d]	NA	NA	180–270	6.1	~95	540	3.5 lb/in.	19.8	MSA
1106BH[e]	NA	NA	180–460	5.8	~100	540	1.5 lb/in.	19.8	MSA
Gelman A[e]	NA	NA	380	9.3	NA	480	NA	18.9	G
E[d]	NA	NA	380	10.0	NA	480	NA	18.9	G
G	NA	NA	810	11.6	NA	480	Gauze reinf.	3.0	G
M	NA	NA	580	10.8	NA	480	NA	6.1	G
H	NA	NA	510	12.7	NA	480	NA	21.7	G
Whatman AGF/A	>1	NA	340	5.3	100	540	230 gm/cm	NA	RA
AGF/B	>1	NA	840	15.0	100	540	560 gm/cm	NA	RA
AGF/D	>1	NA	460	5.5	100	540	100 gm/cm	2.3(20 fpm)	RA
AGF/E	>1	NA	890	15.0	100	540	190 gm/cm	NA	RA
AGF/F	>1	NA	380	6.3	100	540	130 gm/cm	NA	RA
H and V H-93	NA	0.6	460–560	9.3	96–99	540	2.5 lb/in.	NA	H and V
H-94	NA	0.5–3	380	8.2	96–99	480	2.5 lb/in.	NA	H and V
S and S 27	NA	NA	125–180	NA	98	400	NA	NA	S and S
29	NA	NA	200	NA	98	400	NA	NA	S and S

TABLE 6-3 (continued)

Filter designation/ composition	Void size (μm)	Fiber diameter (μm)	Thickness (μm)	Weight per unit area (mg/cm²)	Ash content (%)	Temperature ceiling (°C)	Tensile strength	Head loss at 100 fpm (in. w.g.)	Source[c]
H and V H-70/ cell & asbes	NA	0.1-35	230(9 mil)	8.2	20-25	150	2.5 lb/in.	17	H and V
18 mil, H-70, cell & asbes	NA	0.1-35	460	15.4	20-25	150	4.0 lb/in.	26	H and V
H-64/ cell & asbes	NA	0.1-35	830-1090	22.7	15-20	150	2.0 lb/in.	15	H and V
H-90/ cell & glass	NA	9-35	685	13.4	70	150	3.2 lb/in.	0.41	H and V
H-91/ cell & glass	NA	1.5-35	710	13.5	80	150	3.5 lb/in.	0.89	H and V
N-15/ syn & glass	NA	0.5-15	1270	24.9	15	150	1.0 lb/in.	9.9	H and V
5-G/ syn glas cot	NA	0.5-15	685	14.5	4-6	150	Gauze backed	2.0	H and V
MSA/glas cel	NA	NA	1000	NA	NA	120	NA	NA	MSA
Whatman ACG/A /glas & cell	>1	NA	330	5.5	NA	150	270 gm/cm	0.9(20 fpm)	RA
ACG/B /glas & cell	>1	NA	990	19.5	NA	150	330 gm/cm	2.6(20 fpm)	RA
Miscellaneous fiber filter									
Delbag[f] microsorban/ polystyrene	NA	0.6-0.8	1270	5.4	NA	96	NA	15	G
S and S 1001/ pvc	NA	NA	140	NA	1.4	100	150 kg/cm²	NA	S and S

Membrane filter

Filter designation	Pore size (μm)	Refractive index	Thick-ness (μm)	Weight per unit area (mg/cm²)	Ash content (%)	Temperature ceiling (°C)	Tensile strength	Head loss at 100 fpm (in. w.g.)	Source[c]
Millipore SM	5.0	1.495	170	3.6	<0.0001	125	100 psi	19	M
SS	3.0	1.495	170	3.8	<0.0001	125	150 psi	38	M
WS	3.0	NA	150	4.9	<0.0001	125	NA	~100	M
RA	1.2	1.512	150	4.2	<0.0001	125	300 psi	62	M
AA	0.80	1.510	150	4.7	<0.0001	125	350 psi	91	M
DA	0.67	1.510	150	4.8	<0.0001	125	400 psi	120	M
HA	0.45	1.510	150	4.9	<0.0001	125	450 psi	210	M
WH	0.45	NA	150	5.7	<0.0001	125	NA	~270	M
Gelman AM-1	5.0	NA	200	3.6	NA	125	NA	11	G
AM-3	2.0	NA	200	3.6	NA	125	NA	33	G
AM-4	0.8	NA	200	3.6	NA	125	NA	73	G
AM-5	0.65	NA	200	3.6	NA	125	NA	NA	G
AM-6	0.45	NA	200	5.8	NA	125	NA	NA	G
S and S AF-600	7.5	NA	180-250	6.3-8.7	0.01	100/200[g]	NA	NA	S and S
AF-400	4.0	NA	180-250	6.3-8.7	0.01	100/200	NA	NA	S and S
AF-250	2.0	NA	160-210	5.6-7.3	0.008	100/200	NA	NA	S and S
AF-150	0.85	NA	160-210	5.6-7.3	0.008	100/200	NA	NA	S and S
AF-100	0.70	NA	150 Avg.	5.3 Avg.	0.007	100/200	NA	NA	S and S
AF-50	0.60	NA	135 Avg.	4.7 Avg.	0.006	100/200	NA	NA	S and S
AF-30	0.40	NA	120 Avg.	4.2 Avg.	0.005	100/200	NA	NA	S and S

TABLE 6-3 (concluded)

Filter designation/ composition	Size	Void size (μm)	Temperature ceiling (°C)	Source[c]	Remarks
D 1013/cell	43 X 123 mm	NA	120	WP	Use WP D1012 paper thimble holder
D 1016/glas	2 3/16 X 14 in.	NA	400	WP	Use WP D1015 glass cloth th holder
RA-98/alund	NA	Std.	High	WP	Use WP D1021 alundum thimb holder
RA-360/alund	NA	Fine	High	WP	Use WP D1021 alundum thimb holder
RA-84/alund	NA	Extra fine	High	WP	Use WP D1021 alundum thimb holder
S and S 603/ cellulose	From 6 X 60 mm to 170 X 360 mm	NA	120	S and S	
S and S 703/ sint glass fiber	From 15 X 80 mm to 155 X 180 mm	NA	300	S and S	
Whatman/ cellulose	From 10 X 50 mm to 90 X 200 mm	NA	120	RA	

[a]From Air Sampling Instruments, 3rd ed., Am. Conference of Governmental Industrial Hygienists, Cincinnati, Ohio, 1966, pp. B-2-4 and B-2-5.

[b]NA, Information not available.

[c]Sources: G, Gelman Instrument Company, 106 N. Main St., Chelsea, Michigan; H and V, Hollingsworth and Vose Company, East Walpole, Massachusetts; IPC, The Institute of Paper Chemistry, Appleton, Wisconsin; M, Millipore Filter Company, Bedford, Massachusetts; MSA, Mine Safety Appliances Company, 201 N. Braddock Ave., Pittsburgh, Pennsylvania; RA, H. Reeve Angel and Company, Inc., 9 Bridewell Place, Clifton, New Jersey; S and S, Carl Schiercher and Schuell Company, Keene, New Hampshire; WP, Western Precipitation Div. of Joy Mfg. Company, 1000 W. Ninth Street, Los Angeles, California.

[d]With organic binder. [e]Without organic binder. [f]Soluble in aromatic hydrocarbons. [g]Continuous/

FIG. 6-5. Membrane filter structure (Gelman).

Probably the best characteristic of membrane filters used
for dust sampling is that they can be made transparent for microscopic viewing of the collected dust (see Chapter 9). Membrane
filters arc available in colors (white, green, red, and black),
with or without colored grids (100 squares on the sampling surface of a 2-in. filter) printed on them. The cost for 2-in.-
diameter filters (or 47-mm) runs about $0.15 each for the most
common types up to about $0.30 each or more for special types.
The worst feature (except cost) is that they tend to load up
rapidly because of surface collection.

b. Paper Filters. Paper filters are mats of randomly
oriented cellulose fibers. They have a large amount of void
space and depend heavily upon impaction in order to obtain high
efficiencies for removal of small particles (0.3 um).

Whatman No. 41 filter paper is widely used for air sampling. It is an acid-washed paper and is therefore suitable for solvent extraction of the entrained material. Paper tape sampling, especially for soiling index, is often done with this paper. The cost of 2-in. Whatman No. 41 discs is about $0.01 each.

Paper can be used to temperatures near 300°F. Gravimetric analyses with paper filters require careful equilibration of the filter both before and after sampling.

c. Glass Fiber Filters. Glass fiber filters have much the same structure as paper filters but are made of glass fibers rather than cellulose. A glass fiber filter achieves high efficiencies without requiring impaction velocities.

A glass fiber filter is the filter of choice for suspended particulates and is especially good for solvent extraction. These filters can withstand temperatures to 900°F. Some types have a small amount of organic binder added for strength. The 8 X 10 in. filters used for high-volume sampling of suspended particles cost $0.20 to $0.25 each in lots of 100.

d. Other Filters. Mixtures of fibers may give certain desired filter characteristics. Also, polystyrene and polyvinyl chloride are sometimes used as filter materials. The use of alundum (fused alumina) filters (thimble filters) in stack sampling is much less frequent now that fiber glass is available to withstand most stack temperatures. Nuclepore (General Electric) filters have cylindrical holes, holes that are "punched" by fission fragments then etched to the size desired.

2. Sedimentation Chambers

A sedimentation chamber is simple in design but is very effective for dust sample collection in stacks or places where the concentration is high. The method is most useful for particle size distribution determinations because it avoids the isokinetic and overloading problems commonly encountered with filtration methods. The chamber is normally cubical, from 3 to 18 in. on a side.

Microscope slides are fastened to the top, bottom, and sides of the chamber. The top and bottom should swing on a pivot for quick opening and closing. The slides and their mounts should be recessed into the top and bottom to facilitate the sampling.

The clean sedimentation chamber (prepared in a dust-free space) is put into the area for sampling and oriented with the top-bottom axis along the flow direction. The top and bottom are swung open simultaneously, then rapidly closed after the air has been replaced by the movement of the airstream. The closed chamber is removed to the laboratory where it is left quiescent for a period of several hours or more in order for the particles to be collected on the walls (slides). Particle dynamics considerations show that the small particles are removed by Brownian impaction and the large particles by sedimentation. The 0.2-μm particles are removed slowly by both mechanisms. After the elapsed time for particle collection, the slides are removed for analysis.

EXAMPLE 6-6: What theoretical time should be allowed for the wall collection of 0.2-μm particles ($\rho_p = 2.0$) in a sedimentation chamber that is 20 cm on an edge? For sedimentation?

Maximum distance of travel along diagonal: About 17 cm

$s = \sqrt{[RTt/(3\pi\mu d)]}$ (from Chapter 5)

For $s = 17$ cm; $R = 8.3 \times 10^{7}$ ergs/$^{\circ}$K-mole; $T = 293^{\circ}$K;
   $N_a = 6.02 \times 10^{23}$; $\mu = 1.8 \times 10^{-4}$ poises; $d = 0.2 \times 10^{-4}$
   cm; $t = 3 \times 10^{8}$ sec $= 8 \times 10^{4}$ hr

Maximum distance of travel for sedimentation: 20 cm

$u_t = 0.003\,\rho_p d\,C_c$ (From Eq. 5-26)

For $\rho_p = 2$; $d = 0.2$ μm; $C_c = 1.51$ (see Eq. 5-27);
                          $t = 55,600$ sec $= 15.5$ hr

3. Cyclones

Cyclones are centrifugal collectors which may be used to collect particles above a few microns in diameter. Although cyclones are not employed to a great extent, they are useful for collecting particles greater than some size that depends upon the separation factor (see Chapter 5, Section XI), the residence time, and the distance of travel for removal. Cyclones have sometimes been used for fly ash collection and more often in respirable dust samplers. This sampler is made up of a cyclone which traps those particles that would probably be removed before reaching the deep lung area and a following filter to collect those particles of such a size that they would likely travel deep into the lungs. The cyclone collects about 55% of unit-density 4-μm particles (see Chapter 8). Respirable dust samplers are available (Unico) in sizes from 0.1 to 8.5 cfm at costs from about $25 to $50.

4. Impaction Devices

Several impaction devices have been described in the recent literature. The discussion here is limited to the most popular impaction collectors. The theory and equations for impaction are briefly covered in Chapter 5. The sizing accomplished by impaction is described in Chapter 8.

a. Cascade Impactors. Cascade impactors collect particles in stages according to particle sizes. This permits one to obtain a weight distribution of particle size without the drudgery of counting. The cascade impactor is made up of a series of stages (four to six or more); each stage has holes or slots smaller than the preceding stage; the successively higher stage velocities result in the impaction of smaller and smaller particles.

The principal types of cascade impactors in general use are the Casella (May), the Unico (Lippmann), and the Andersen.

The Casella employs a flow rate of 17.5 liters/min. It has four
stages set at right angles to one another. The stage consists
of a slot to give the desired velocity and a backing plate to hold
a glass slide cover at a fixed distance from the jet. The Casella
cascade impactor has been modified by some workers to prevent
air leakage at the joints; the marketed version of the original
has cast threads which are unsatisfactory for obtaining leak-
proof joints. A filter is usually added to follow the fourth stage.

The Unico cascade impactor is popular and offers several
advantages over the Casella. Collection is on standard microscope slides (1 X 3 X 1/16 in.) as shown in Fig. 6-6. The
slides can be seen during sampling, and the collection position
can be changed by moving the slides to several different notches
of a ratchet device. This avoids overloading, which is a common
complaint in regard to cascade impactors. The Unico impactor
has been calibrated at various air flows.

The Andersen sampler has 400 holes to pass 1 cfm through
in each of six stages. The holes become progressively smaller
from 0.0465 in. ⌀ (3.54 fps) in stage one to 0.0100 in. ⌀ (76.4 fps)
in stage six. The sampler is available in two versions--the
bacteriological model collects particles on agar in petri dishes
and the other model collects particles on stainless steel or glass
plates.

The prices for the cascade impactors are approximately
$100 for the Casella, $150 for the Unico, and several hundred
dollars for the Andersen (which has its own pump).

Cascade impactors collect particles above a few tenths of
a micrometer in diameter. Their efficiencies and calibration are
described in Chapter 8, Section II, D, 1, d).

b. Impingers. Impingers collect particles by impaction
on the surface below the inlet nozzle and to some extent by

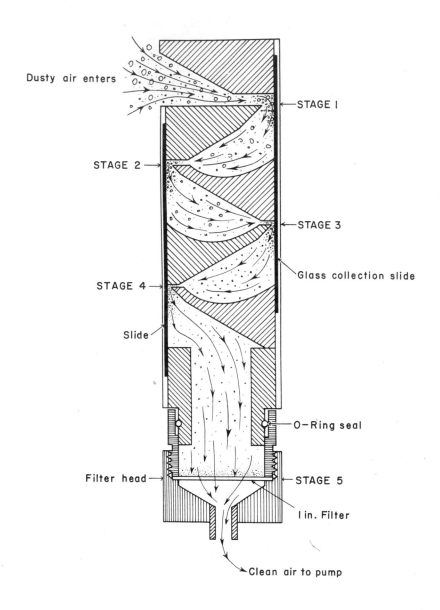

FIG. 6-6. Cascade impactor (Unico).

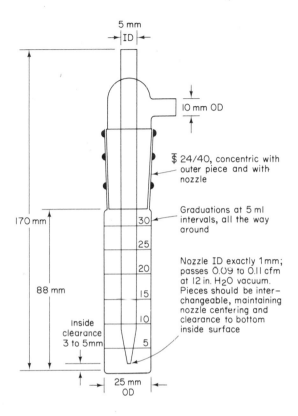

5 mm
→|ID|←

10 mm OD

⊊ 24/40, concentric with
outer piece and with
nozzle

Graduations at 5 ml
intervals, all the way
around

Nozzle ID exactly 1mm;
passes 0.09 to 0.11 cfm
at 12 in. H$_2$O vacuum.
Pieces should be inter-
changeable, maintaining
nozzle centering and
clearance to bottom
inside surface

170 mm   30
                25
                20
88 mm   15
                10
Inside   5
clearance
3 to 5mm

25 mm
OD

FIG. 6-7. Midget impinger. From B. E. Saltzmann et al.,
Selected Methods for the Measurement of Air Pollutants, U.S.P.
H.S. Publ. No. 999-AP-11, Cincinnati, Ohio, May 1965.

scrubbing. Two basic sizes of impingers are used in particulate
sampling--the midget impinger for a flow of 0.1 cfm and the
Greenburg-Smith impinger for 1 cfm. Figure 6-7 shows a midget
impinger and its dimensions. The midget impinger is used much
more often than the larger size. A microimpinger (0.02-0.025
cfm) with a solution capacity of 2 ml is marketed by Unico.
Impingers collect about 85% of 1-μm particles, but the efficiency
falls off rapidly for particles smaller than 1 μm. Impingers cost

from about $7 to $20 and are available with fritted glass bubblers
for gas absorption.

    c. Other Impaction Devices. Impaction devices other than
cascade impactors fall into two general groups--those that have
mechanical movement and those that depend upon the wind to
carry out the impaction. The most widely used are the rotating
rod, bar, slide, or disk, which have motor-driven rotation of the
collecting surface. The most frequent application of such samplers is for the collection of pollen and airborne microorganisms.
The samplers have also been used for tracer studies. The rotating collector is usually adhesive coated and turns at a maximum
tangential velocity of about 25 mph (the velocity should be
greater than the wind speed). The rotating rod has variable
velocity as the distance from the center of rotation increases,
which gives some gradation of collected size. Other mechanical
impaction devices include the annular impactor, centripetal
impactor, slit sampler (bacterial), and jet filter.

    Wind impaction is the mode of collection on an adhesive-
coated drum designed by Gruber. The direction of the wind that
carried the particles is determined by the position on the drum.

5. Electrostatic Precipitation

    An electrostatic precipitator puts a charge (-) on particles
that pass through the corona leakage, causing the particles to
migrate to the oppositely charged electrode (+) where they are
collected. The collector is suitable for either liquid or solid
particles. It has a high collection efficiency on usual samples
and does not crush or break the particles. The corona leakage
is a rather efficient ozone generator and the instrument cannot
be used in explosive atmospheres.

    A popular electrostatic precipitator air sampler (Mine
Safety Appliances) has a number of aluminum tubes (1 1/2 in.

o.d. X 7 1/2 in. long) which are used as collectors. One of
the tubes serves as the anode, and a wire at the center of the
cylinder is the cathode. The operating potential is 12-16 kV.
Particles may be collected directly in the tube or the tube may
be lined with paper or a membrane or other material. The tubes
have a high finish and are easily washed. The instrument in
its original form weighs more than 30 lb and costs about $600.
Another company markets a 6 1/2-lb instrument which gives considerably more portability. Other electrostatic precipitators
have been built to collect the sample directly on a slide.

The electrostatic precipitator has been used to collect a
sample on a grid for electron microscopy. There has been
some debate as to whether or not the sample is laid down uniformly over the Formvar coating.

Recent investigations have shown that a uniform, repre-
sentative deposit is not laid down with the normal electrostatic
precipitator. Liu has developed a two-stage, pulsing electrostatic precipitator for overcoming this difficulty.

6. Thermal Precipitators

A thermal precipitator collects particles by the thermal
force on the particle in the thermal gradient, a force from the
hot to the cold. A thin wire is heated and air is passed between
the wire and a colder plate or between two plates (glass slide
or cover glass on metal backing). The deposit is not laid down
uniformly. In order to obtain uniform deposition, the slide is
given a reciprocating motion.

The thermal precipitator is efficient for small particles,
collects particles without crushing, and gives a finished sample
(collected on slide). It has a low efficiency for large particles
(>10 μm) and a very low flow rate, which combine to limit its
applications. The cost is over $600.

7. Other Particle Collectors

Particle collectors not yet described include combination collector-analyzers such as jet dust counters and some of the simple means such as dustfall sampling directly on a slide and collection of settled dust from a surface by sticky tape or a camel's hair brush. Combination collector-analyzer methods are discussed in Chapter 8. The simple means do not really need describing. One precaution is to use a tape that does not change color with time if it is to be placed on a slide with its dust, or one that is soluble in an available solvent if it is to be dissolved for analysis.

8. Absorbers

As a rule an air sample is bubbled through an absorbing solution to collect a gaseous fraction. Sometimes the sample is put into a flask with the absorber; then the stoppered flask is shaken vigorously to allow long intimate contact between the sample and the absorber if the reaction is slow.

Absorption of a gaseous component from an air sample is usually accomplished by a midget impinger. When the efficiency of the removal is low, a second and sometimes a third impinger may be used in series. Fritted glass bubblers are normally used to give smaller bubbles than the single jet for higher efficiency of absorption. Such bubblers often become fouled with particles in a dirty atmosphere; for this reason many investigators do not use fritted bubblers at all. Evaporation of absorber frequently limits the sampling period to 15-30 min, or requires the addition of makeup absorber during sampling. In some cases the air can be saturated with a water impinger before passing through the absorber solution.

9. Adsorbers

Adsorbers are used to collect many dilute organic gaseous pollutants, especially odors. Much larger samples can be put through an adsorber bed than through an absorber. Activated carbon is nearly always the adsorber matrix because it can be used in the presence of moisture, whereas the other adsorbents are polar in nature and can be used only in a dry airstream.

The adsorbed material is stripped from the adsorbent (carbon) by heat and vacuum, steam, or solvent (chloroform, ether, and so on). Adsorber beds are frequently made by filling a glass cylinder with granular activated carbon held in place in the cylinder with glass fiber plugs. Gas mask canisters have been used with high-volume samplers. Such use gives very little contact time between the air and the carbon. Activated carbon has been impregnated with chemicals to collect certain contaminants, especially radioiodine.

10. Condensation

Condensation of fractions from an air sample is often carried out using a series or train of traps cooled to various temperatures. The condensation of any fraction at very cold temperatures requires the removal of the water vapor before the cold trap. A train may have the traps cooled by: cracked ice ($0^{o}C$), or even cool water; cracked ice with salt($-18^{o}C$); dry ice ($-78^{o}C$); liquid oxygen ($-183^{o}C$); and liquid nitrogen ($-196^{o}C$). To prevent explosions care must be exercised in the use of liquid oxygen.

D. Automated Samplers

Most of the automation of air samplers involves sampling time, that is, when and how long to sample. When and how long may mean concurrent with some emission or at a given time. Automation is sometimes needed but scarcely ever made on the maintenance of isokinetic conditions for ambient particulate sampling--to keep the probe pointing into the wind and to adjust the probe velocity in accordance with wind changes. Isokinetic conditions are important only for large particles (see Section IV, D).

Sequential samplers are available for changing from one port to another at given intervals, with or without a period of no sampling at the change. Typically, the time for sampling a given port (collector) is adjustable up to 2-4 hr, and the pause between ports adjustable up to the same periods. A paper tape sampler moves the filter tape to a new spot on the tape rather than changing to a new filter holder.

Samplers that collect airborne particles by impaction, such as rotating rods or slides, may sample intermittently (e.g., 10 sec in 15 min) to prevent overloading. Normally, periodic sampling is made for correlating the sampling results with time or with events that occur at certain times, such as 8 a.m. and 5 p.m. traffic peaks.

A sampler used to determine airborne bacteria with the time of day has an impaction slit that is moved over an agar surface by clockworks. A device similar to the bacterial impaction sampler but used for dirtiness with wind direction has a slit moved over the collector surface by a wind vane.

If it is desired to compare the dust removed from the atmosphere by sedimentation with that removed by rainout or washout, a switching mechanism activated by moisture may be used to cover and uncover sampler jars. Similar devices may be temperature, concentration, or pressure controlled for special sampling determinations. The last-mentioned has long been used for meteorological soundings.

IV. TECHNIQUES

The subject of sampling techniques is as broad as the imagination or ingenuity of the person doing the sampling. It is impossible to describe all sampling situations and their attendant properties. This discussion is intended to be rather general and to point out a few types of attack that have been used on some air pollution sampling problems.

A. Calibration

In order for test results to have significance, considerable calibration must be carried out on all steps of the test from the sampling to the analysis. The volume of sample, the efficiency of collection, and the efficiency of analysis must be calibrated. In many cases the calibration of the efficiencies of collection and analysis are combined.

1. Metering Devices

The spirometer is the primary standard metering device; therefore all other metering devices should be calibrated against a spirometer. The calibration should be done at varying flow rates because the flow rate often affects the meter reading. Calibrations are easier to perform with a large-capacity spirometer (5 ft^3, for example) than with one having a smaller capacity. Stopwatch measurements can be made over easily measurable time periods even when rather large flow rates are used. Measurements should be started after some flow has moved through the meter in order to remove any slack in the system and to have a uniform flow rate during the timed period.

2. Instrument Flow Rates

High-volume air samplers have been calibrated in a number of different ways. The most satisfactory method is again the use of a spirometer; however, calibrations are often carried out with dry test meters. Measurements have also been made by hot-wire anemometers in a duct section fastened to the exhaust of the sampler; this is a rapid method and merits use when large numbers of samplers are to be calibrated. Much of the work is done with an orifice and manometers; for example, the NASN does field calibrations with an orifice outfit (General Metal Works, Inc.).

Very low-head pumps, such as electrostatic precipitators, are limited in the number of applicable calibration methods.

Tebbens used a dilution tank for such measurements. A large
tank filled with a pure gas or vapor, carbon dioxide, nitrogen,
carbon tetrachloride, or other, can be exhausted by a low-head
pump. If complete mixing of the air coming into the tank is as-
sured by use of a fan, the concentration of the gaseous material
in the tank changes as a first-order equation.

$$dC/dt = -kC, \qquad\qquad\qquad (6-4)$$

or integrated

$$C = C_0 e^{-kt},$$

where C_0 = original concentration of the gas (t = 0), C = concentration at any later time t, and k = constant of decay = multiple
of tank volume removed in time t. When C/C_0 = 0.37; that is,
kt = 1, the tank volume has been displaced. A plot of C/C_0 on
a log scale versus t on an arithmetic scale gives a straight line
with a slope of -k, and it is appropriate to mention the half-life
(0.693/k), the time required for the concentration to be reduced
by one-half.

3. Collector Efficiencies

Collector efficiencies can often be tacitly assumed to equal
100%. This is true for most filters in most collection situations,
especially membrane filters.

If gaseous absorption must be calibrated, it is necessary
to have a known source or standard. A chamber can be prepared
by injecting small amounts, perhaps with a hypodermic syringe,
of the gas to be calibrated into a much larger volume and mixing.
Mylar bags or other flexible containers are probably the easiest
to use. Permeation tubes are made by putting the liquid material
into a permeable tube and closing the ends. These tubes are
available for a number of gaseous pollutants. The permeation
tube is placed in a carefully controlled flow of air which is free
of the material to be tested. After carefully mixing the air and
the material picked up in its passage by the permeation tube, a

standard stream of gas is obtained. The concentration must be
known and is usually calculated from a specified rate of leakage
of the gas from the tube as determined by carefully weighing the
tube over long time intervals and the carefully determined flow
rate of the dilution air.

The efficiency of collection for a particulate sampler is
most often checked for a standard particle distribution such as
DOP smoke or methylene blue or a monodisperse (constant size)
aerosol such as the polystyrene latex particles now available
(Dow). This subject is covered in more detail in Chapter 8.

A calibration technique sometimes used involves placing
two collectors in series and assuming that the efficiency of the
second collector is equal to that of the first, a not too gross
assumption for monodisperse aerosols or gases; then

$$\eta = 100 \ (a - b)/a \ , \tag{6-5}$$

where η = efficiency of collection (%), a = amount collected by
first collector, and b = amount stopped by second collector.

B. In Toto versus Concentration Sampling

Most air sampling is accomplished by passing air through
a collector to remove the constituent of interest through filtration,
gravity or inertial means, or electrical or thermal precipitation
for particulates, or by absorption or adsorption of gaseous materials. There are times, however, when it is desired to sample the
total mixture.

In toto sampling may be done by on-the-spot analysis techniques such as putting an airstream through an IR spectrophotometer to determine carbon monoxide or by sample removal for later
analysis, such as filling a Mylar bag with a sample for laboratory
analysis by IR spectrophotometry or GC. This method of sampling
also has an advantage in that it permits the collection of large
samples in short periods. For example, sulfur dioxide absorption

in a midget impinger may take at least 10 min to bubble through
the desired volume, but a bag could be filled in much less than
1 min and the sample passed through an impinger later. Puff
releases can be sampled in this manner to obtain short-term
transients in concentrations.

Concentration of the sampled constituent at the time of
sampling provides a means of sampling volumes that run into
hundreds or even thousands of cubic meters. This method gives
a nearly pure sample; that is, the material for analysis is in a
much more concentrated form and requires less sensitivity of
analysis than the very dilute concentrations found in the air.

C. Sample Storage

Some samples can be stored indefinitely without undergoing
changes in their nature, while others require immediate analysis.
Quartz particles, once they are deposited on a slide or a filter,
are not changed with time as long as they are covered; however,
the ozone fraction of a sample has a short lifetime and requires
early analysis. Some nonconservative materials can be removed
from the air and chemically fixed for making later determinations.

Checks should be made to assure that errors caused by
storage are known and can be corrected for. Storage of gaseous
samples in Mylar bags can be successful if the rate of decay
from chemical reaction, adsorption on the bag, or interchange
with the surroundings is known. Using another investigator's
rate of decay is not a good practice because it is hardly likely
that his sample has all of the same characteristics as yours.

D. Isokinetic Sampling

The inertial bias introduced in using a sampling velocity
different from the airstream velocity can be a serious source of
error in the collection of particles. Simple logic shows that a
sampler pointing into an airstream with a sampling velocity

greater than that of the airstream bends streamlines into the
opening, and the small particles tend to follow the streamlines
while the large particles from the outer streams tend to cross the
streamlines and pass by without entering the probe; therefore a
bias exists in favor of collecting the small particles and a dis-
proportionate fraction of small particles are collected. If the
sampler is pointed downstream or cross stream, too few large
particles are collected in any case.

Watson used the impaction parameter (ψ) concept from
aerosol mechanics to fit data collected with spores 32 and 4 μm
in diameter. An adaptation of his formula is

$$C/C_0 = U_0/U \{1 + [1.23 - \sqrt{(0.053 + 0.366\,\psi)}][\sqrt{(U/U_0)} - 1]^2\},$$

(6-6)

where C = measured concentration of particles with diameter d
(cm^{-3}), C_0 = true concentration of particles with diameter d (cm^{-3}),
U = stream velocity (cm/sec), U_0 = mean probe velocity (cm/sec),
and ψ = impaction parameter (see Chapter 5).

Figure 6-8 shows a family of curves calculated from Eq.
6-5 for various particle diameters. When correcting for anisokinetic conditions, it is most important that cognizance be taken
of the tremendous influence of a single large particle on the mass
distribution of the sampled particles.

Most particulate sampling in stacks and ducts is carried
out isokinetically; however, ambient air sampling is seldom
done in this manner. Often such sampling is made without even
facing the sampler into the wind. This may not be as serious as
it seems because most of the large particles have settled out of
the air. Suspended particulates are often measured in a shelter
with the sampler pointed upward. Brookhaven National Laboratory
built an isokinetic ambient sampler--a vane keeps the nozzle
pointed into the wind and an anemometer adjusts the sampling
rate to equalize the nozzle velocity with the wind velocity.

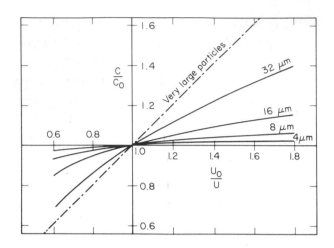

FIG. 6-8. Plot of equation 6-5. From H. H. Watson, "Errors Due to Anisokinetic Sampling of Aerosols, Am. Industrial Hyg. Assoc. Quart., 15:1, 21-25, March 1954.

EXAMPLE 6-7: What face velocity does an 8 X 10 in. filter have when the sampling rate is 70 cfm? Assume full face utilization.

$$70 \text{ cfm}/(8 \times 10/144 \text{ ft}^2) = 126 \text{ fpm} = 1.43 \text{ mph}$$

E. Duct and Stack Sampling

In air pollution work, stack sampling is done to determine the pollutant emissions being discharged by the stack and the probable rise. The parameters normally measured are velocity and area, concentration of pollutants, size of particles, temperature, and dew point or water content. Duct sampling is for either of two reasons; namely, it is easier to sample in the duct than in the stack, or it is desired to determine the efficiency of an air cleaner by sampling before and after the cleaner.

Sometimes the average duct or stack velocity is taken to be 0.9 times the velocity at the center line. This is not a bad approximation after a long run (10X diameter) of straight duct; however, because of the usual lack of uniformity of velocity and particulate concentration across the entire cross section of the duct or stack, a sample traverse is normally laid out prior to sampling and rigorously adhered to during sampling. The sampling points are usually located such that all of the points represent equal fractions of the total area. Figure 6-9 shows layouts for rectangular and circular cross sections. The radii for the division circles are calculated by

$$r_k = \sqrt{(kD^2/4n)} = D/2 \sqrt{k/n}, \tag{6-7}$$

where r_k = radius of the kth subarea, k = subarea index, D = duct diameter, and n = number of subareas (circle plus annular rings).

EXAMPLE 6-8: Determine the locations of sampling points for a 48-in. circular duct. It has been decided on the basis of the particular layout and situation that 16 points will suffice.

D = 48 in.   n = 4

$r_1 = 24\sqrt{1/4} = 12$ in. $r_2 = 24\sqrt{1/2} = 17$ in.

$r_3 = 24\sqrt{3/4} = 20.8$ in. $r_4 = 24\sqrt{1} = 24$ in.

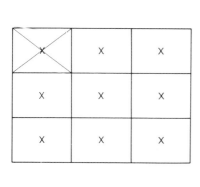

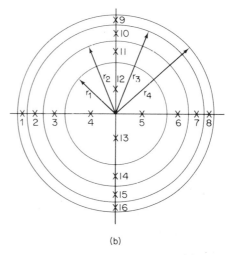

(a)   (b)

FIG. 6-9. Traversing points for ducts or stacks. (a) Rectangular. (b) Circular.

Distances of sampling points from center:

1* $r_1/2 = 6$ in.   2 $(r_1 + r_2)/2 = 14.5$ in.

3    $(r_2 + r_3)/2 = 18.9$ in.   4 $(r_3 + r_4)/2 = 22.4$ in.

*Perhaps the location of the point in the sector should be at $\sqrt{2}/2\, r_1 = 8\ 1/2$ in., but this procedure is not justified.

Distances from left and top walls (or west and north):

1 and 9 $24 - 22.4 = 1.6$ or 1 5/8 in. (to be measured)

2 and 10 $24 - 18.9 = 5.1$ or 5 1/8 in.

3 and 11 $24 - 14.5 = 9.5$ or 9 1/2 in.

4 and 12 $24 - 6 = 18$ in.

5 and 13 $24 + 6 = 30$ in.

6 and 14 $24 + 14.5 = 38.5$ or 38 1/2 in.

7 and 15 $24 + 18.9 = 42.9$ or 42 7/8 in.

8 and 16 $24 + 22.4 = 46.4$ or 46 3/8 in.

The velocities measured for the points are then averaged. The pressures read with a pitot tube should not be averaged because velocity is proportional to the square root of the pressure; therefore the square of the average of the square roots of the pressure readings should be used .

V. SUMMARY

A good job of air sampling is very likely to be properly planned beforehand and carefully carried out. Good sampling is often costly, but poor sampling may be even costlier. One major equipment manufacturer discovered this when it put in standard cyclones to collect particles with diameters less than 3 μm.

Before beginning to sample, one needs to know what, where, when, how long, how much, and why. Each item plays a vital role in the success of the undertaking.

In general, sampling requires an air mover, a calibrated flowmeter, and a collector. The most popular outfit probably consists of a carbon vane pump driven by an induction motor, a rotameter, and a midget impinger. This assembly gives maximum flexibility of sampling for gaseous and particulate pollutants.

Air movers or pumps should be matched to the job, for both flow rate and operating head. Flow rates range from 20 cm^3/min for thermal precipitation to 2 m^3/min for a high-volume sampler, and heads range from about 0.1 in. w.g. for an electrostatic precipitator to 25 in. mercury for small-pore membrane filters. Sometimes, particularly in research, the filter is used on the pressure side at heads up to several atmospheres. Many types of pumps are used in air sampling. These include piston, lobe, eccentric vane, diaphragm, and centrifugal. Actually, what is required is a vacuum source; car manifolds, aspirators, Freon-powered ejectors, evacuated flasks and tanks, water displacement, and other air movers have been used.

Meters should be calibrated against a primary standard, a spirometer. Laboratory meters can be large and bulky, such as the dry test and wet test meters, but field instruments should be small and portable such as rotameters, orifices, pitot tubes, or deflecting vanes or other anemometers. The flow rate is normally taken as the average of the initial and final readings. The variation from straight-line change is of less significance than other practices in approximating the flow.

Particle collectors remove particles by sedimentation, centrifugal force, impaction, electrostatic precipitation, thermal precipitation, or diffusion. Filters work principally by impaction or interception on particles larger than several tenths of a micrometer and by diffusion on those smaller than 0.1 μm. Interception or straining plays a minor role in the small particle collection, but it can assure collection of particles greater than the pore size of the filter. A filter with a large pore size may collect virtually all of the particles orders of magnitude smaller in diameter than the pore size and will do so at a much higher flow rate than a filter with the pore size less than the diameter of the particles.

Gaseous materials can be absorbed in a collecting liquid, adsorbed on a solid matrix, or condensed by cooling in traps. Materials measured in the parts-per-million range are usually absorbed in an impinger of suitable absorbent and those found in parts per billion or less are adsorbed or condensed, especially odorous organic substances.

Passive air sampling may consist of putting a material in the ambient air and depending upon the wind movement and diffusion to move the air by the sampler. This type of test is used for corrosion, sulfation rate, ozone, and dustfall measurements.

PROBLEMS

1. Particulate sampling near a lime plant shows $1250 \, \mu g/m^3$ of suspended particulates during a 3-hr period. Assuming rapid, short-term mixing, calculate the probable 24-hr concentration and the 30-min concentration.

2. How much air must be sampled to obtain several micrograms of phenol where the ambient concentration is about $0.5 \, ppm_v$

3. What method of collection should be used for problem 2? What kind of sampler should be selected? What is the probable sampling time required?

4. List the equipment and procedures best suited for the determination of particulate sulfates in your city. Use statistical calculations at the 90% confidence limit and a 25% standard deviation. The maximum error should be 25%.

5. What pitot tube head will be read for a stack gas (mostly air) that has a velocity of 10 m/sec and a temperature of $350^\circ F$ if red oil (s.g., 0.6) is used in the manometer?

6. What diameter critical orifice is needed to obtain a flow rate of 0.1 cfm at NTP? Compare this size with the inlet jet of a midget impinger (1 mm).

7. If a concentration of $2000 \, \mu g/m^3$ is reduced to $740 \, \mu g/m^3$ in a completely mixed dilution tank over a period of 24 min, what is the k value in hr^{-1}? What is the flow rate if the volume of the tank is $20 \, ft^3$?

8. If two filters are placed in series to sample airborne radioactivity and the filters show net counts of 720 ± 10 cpm
   and 20 ± 4 cpm, respectively, what are the efficiencies
   of the filters when the particles are assumed monodisperse?

9. Determine the locations for 16-point sampling in a 3 X 4 ft
   rectangular duct.

BIBLIOGRAPHY

Air Sampling Instruments, 3rd ed., American Conference of
    Governmental Industrial Hygienists, Cincinnati, Ohio,
    1966.

A. P. Altshuller, "Air Pollution," Anal. Chem., 39:5, 10R–21R,
    April 1967.

C. N. Davies, M. Aylward, and D. Leacey, "Impingement of
    Dust from Air Jets," Arch. Industrial Hyg. Occupational
    Med., 4, 354–397, 1951.

R. I. Larsen, "A New Mathematical Model of Air Pollutant Con-
    centration Averaging Time and Frequency," J. Air Pollution
    Control Assoc., 19:1, 24–30, January 1969.

R. I. Larsen, "Future Air Quality Standards and Industrial Con-
    trol Requirements," Proc. Third Natl. Conf. Air Pollution,
    Washington, D. C., 199–204, December 12–14, 1966.

M. Lippmann, "A Compact Cascade Impactor for Field Survey
    Sampling," Am. Industrial Hyg. Assoc. J., 22, 348–353,
    October 1951.

B. Y. H. Liu and A. C. Verma, "A Pulse-Precipitating Electro-
    static Aerosol Sampler," Anal. Chem., 40:4, 843–846,
    April 1968.

D. R. Lynam, J. O. Pierce, and J. Cholak, "Calibration of the
    High-Volume Air Sampler," Am. Industrial Hyg. Assoc. J.,
    29:6, 83–88, January–February 1969.

R. O. McCaldin and E. R. Hendrickson, "Use of a Gas Chamber
    for Testing Air Samplers," Am. Industrial Hyg. Assoc. J.,
    20:6, 509–513, December 1959.

P. L. Magill and R. D. Cadle, "Sampling Procedures and Meas-
    uring Equipment," Air Pollution Abatement Manual (C. A.
    Gosline, Ed.), Chapter 6 and Supplement, Manufacturing
    Chemists' Assoc., Washington, D. C., 1952 and 1956.

"Polystyrene Latex Aerosols Used to Determine Collection Ef-
    ficiency of Filter Paper," Hazards Control, I:10, 2–3,
    October 1961. (Livermore Radiation Laboratory Publ.)

S. Posner, "Air Sampling Filter Paper Retention Studies using
     Solid Particles," Proc. 7th AEC Air Cleaning Conf.,
     Brookhaven Natl. Laboratory, Upton, New York, October
     1961.

I. A. Singer, K. Imai, and R. G. del Campo, "Peak to Mean
     Pollutant Concentration Ratios for Various Terrain and
     Vegetation Cover," J. Air Pollution Control Assoc., 13:1,
     40-42, January 1963.

N. G. Stewart, H. J. Gale, and R. N. Crooks, "The Atmospheric
     Diffusion of Gases Discharged from the Chimney of the
     Harwell Reactor BEPO," Intern. J. Air Pollution, 1:1, 31-
     43, January 1958.

B. D. Tebbens and D. M. Keagy, "Flow Calibration of High-
     Volume Samplers," Am. Industrial Hyg. Assoc. Quart.,
     17:3, 327-332, September 1956.

H. H. Watson, "Errors Due to Anisokinetic Sampling of Aerosols,"
     Am. Industrial Hyg. Assoc. Quart., 15:1, 21-25, March
     1954.

Chapter 7

ANALYTICAL--THEORY, EQUIPMENT,
PRINCIPLES, AND TECHNIQUES

I. INTRODUCTION

Proper analysis must follow correct sampling, concurrent
in some cases, in order to serve the intended purpose. Successful analysis requires prior planning, selection of method, careful
procedures, and presentation of the collected data. Analytical
instrumentation is rapidly becoming more complex and expensive.
The principles involved in equipment used for and techniques
developed for analysis are discussed in this chapter; physical
analysis of particulates and specific test methods are treated
in subsequent chapters.

II. THEORY

A. Planning for Analyses

Proper analysis, just as proper sampling, requires certain
prior considerations to assure success. Not only does the
analyst need to be cognizant of the test methodology and how to
carry it out, but he must also know the limitations of his analysis,
the precision and accuracy, and the interferences to be expected.
He should know and consider the cost of the analyses in making
his choice of method and/or equipment.

B. Purpose of Data Collection

The purpose of data collection was discussed sufficiently
in Chapter 6; however, reiteration of the subject here may point
to the importance which attaches to this topic. The analyst

247

must know the purpose of data collection in order to provide
information that adequately serves the needs with the minimum
expenditure of time and money.

C. Precision, Accuracy, and Interferences

There are no exact measurements, but measurements should
be made with enough exactitude to satisfy the need for the
measurement. There are many cases in which much more precise
data than are needed for the task at hand require only a modest
increase over the minimum effort. Later use may develop for
the data if this additional expenditure of effort is made. Also,
good practice often dictates that the work not be done sloppily
regardless of the needed precision.

Accuracy of measurements depends upon, but not solely,
careful calibration of the process. The subject of calibration
techniques is discussed later in this chapter. Interferences
often ruin the accuracy of gathered data, and careful calibration
will find some of these. One of the hazards of using old data
lies in assuming that the data were collected in the manner
currently employed, one that removes certain interferences.
For example, early NASN data for sulfur dioxide contained
interference from nitrogen oxides. The point is that precision
does not necessarily indicate accuracy.

D. Limits of Detection and Sensitivities

There is a minimum amount of material that can be detected
by an analytical method. This minimum may be set by the accuracy from a titration or other laboratory technique, or it may be
a limitation of the equipment used, especially the noise in
the equipment.

The precision and accuracy of a test can be determined
by running blanks and known samples of the material. From the
size of the errors and error propagation theory, the total error

likely in a measurement (under test conditions) can be calculated.

Statistical analysis can be used to find the minimum sample signal needed when the noise level of an analytical instrument is known. Instrument noise (N) can be found from the fluctuations in the recorder trace when an unvarying sample (not a radioactive sample, for example) is in the analysis position or cell (see Chapter 4, Section XIII). The limiting detectable concentration (according to St. John) is found from

$$(S/N)_{lim} = t\sqrt{2/n}, \tag{7-1}$$

where $(S/N)_{lim}$ = minimum signal-to-noise ratio for confidence limit specified in the value of t, t = student's t from statistical tables, and n = number of pairs of sample-background measurements. The signal strength (S) is the average strength after the equipment reaches a steady condition (~ 5 X response time).

The analytical sensitivities of various methods are shown in Table 7-1. These values are practical numbers believed to

TABLE 7-1

Approximate Sensitivities of Analytical Methods

Method	Sensitivity
Neutron activation	0.1 pg
Mass spectrometer	Nanograms
Gas chromatograph	
Electron capture	Picograms
Flame ionization	Picograms
Thermal conductivity	Nanograms
Spectrophotometers (IR, UV, visible)	Micrograms
Wet chemistry	Micrograms
Gravimetric	Micrograms to milligrams

be generally acceptable to investigators in the field. They
are intended to be orders of magnitude and are being lowered by
various new developments in technique and equipment.

E. Response Time

Any analysis carried on concurrently with sampling that
moves an airstream through an analyzer has a response time.
The response time is defined as the delay between the time of
a concentration change and the time the analytical response to
that change has reached within several percent (maybe 5 or 10%)
of its total. Response times become significant when the
desired averaging time is quite short, which is the usual case
for odorous materials, especially sulfur dioxide, and research
into dispersion mechanisms.

The response time is necessarily long enough for the volume
of the lead-in tube and the analytical cell to be filled at the flow
rate sampled. This condition applies only to plug flow; mixed
flow takes longer. Larsen and co-workers state that the response
time and other related times they define (time constant, lag
time, and delay time) may be expressed by

$$t = \frac{k V}{Q},\tag{7-2}$$

where t = response time (or other), V = sensor volume, Q =
volumetric flow rate, and k = a characteristic parameter for the
system. They also recommend that the true concentration be
calculated from

$$C = C_0 + p \frac{dC_0}{dt},\tag{7-3}$$

where C = true concentration, C_0 = observed concentration,
p = time constant of system [time to 63% of maximum response
or $1 - (1/e)$], dC_0/dt = rate of change of C_0. The C values may
be plotted versus a time scale shifted by the selected time constant, perhaps one-half of the averaging time desired, to obtain
a graph of the true concentrations existing at given times. The

time constants (in minutes) for some CAMP instruments are carbon monoxide, 0.04; nitric oxide and nitrogen dioxide, 10; and sulfur dioxide, 3.5.

F. Selection of Method

Several factors should be considered in selecting an analytical method and the equipment for that method. Items to be considered are the levels to be measured and the sensitivity, and specificity, reliability, and extent of usage of the method or equipment. Other important factors are the amount of sample preparation, time required for analysis, frequency of analysis, probable cost per analysis, and maintenance or service required for the equipment. The method or equipment that will do the best job for the application at hand should be chosen without reference to "what everybody else has."

G. Presentation of Results

Efforts should be made to report analytical data in a manner that conveys the information gained--all of it and no more. Consideration of the information may well indicate the manner of presentation.

Sometimes it is desired to present the measurements without any attempt to show trends, correlations, or patterns; this can be done by giving the measurements in a series for a single parameter or in tabular form when two or more entries represent different properties of the same sample or sampling time. Space saving in data presentation is seldom worthwhile if it is done at the expense of readability.

Trends and correlations can normally be shown best by graphs that show one variable plotted against the other. For three variates a family of curves on a regular two-variate plot often serves very well, each curve in the family representing a single value of the third variate to accompany the values on the ordinate and the abscissa. In some cases a three-dimensional plot presents the data in the most desirable manner. If coordinate scales can be selected to give straight lines, the information is transferred to the reader with a minimum of effort on his part, and that is the purpose of recording the data. Efforts to obtain straight-line plots should not cause the analyst to resort to obscure scales which few people understand unless there are good reasons to do so. Some types of graphs have become more or less standardized; papers are available to plot them on, and they should be used whenever possible.

The preservation of directional orientation is achieved in the wind rose and the pollution rose patterned after the wind rose (see Fig. 7-1). The wind rose shows velocities, their durations, and their directions in a compact form.

The statistical limits on listed measurements should be shown. For plotted points the limits may be indicated by wings on individual points (see Fig. 10-3a), or by containing curves (see Fig. 10-3b). Unless otherwise stated, such limits are assumed to be one standard deviation ($\pm \sigma$ or $\overset{x}{\div} \sigma_g$) away from the point or curve.

None of the techniques for using statistics to deceive should be employed for technical data, for example, the use of volumetric containers whose heights are the real measures on a bar graph or a statement of the error in an instrument as less than 5% of full scale; the former is misleading because the reader thinks of volumes, not heights, and the latter because the 5% error is remembered but 5% of full scale may well mean more than 100% error in low-range measurements.

When a line or curve is used to represent data, all of the points should be shown unless the number of points is quite prohibitive; then a randomly selected portion of the points may be used. Omission of data points that do not follow the desired hypothesis is indeed hazardous; often the apparently erratic data are reproducible, and there have been instances in which the blip away from the curve was the most valuable point in the data. Chauvenet's criterion is sometimes used to justify the exclusion of an errant data point (see Chapter 3).

By drawing conclusions from certain aspects of data presented, one may gain additional insight into the nature of the data presented. For example, the arithmetic averages given for the parameters in Table 1-3 can hardly refer to a symmetrical

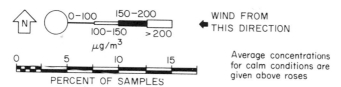

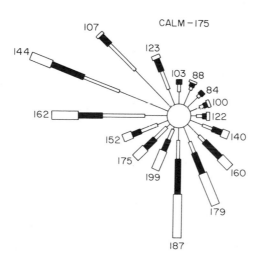

FIG. 7-1. Pollution rose. From J. R. Farmer and J. D. Williams, _Interstate Air Pollution Study: Phase II Project Report--III. Air Quality Measurements_, U.S.P.H.S., Cincinnati, Ohio, December 1966.

(normally distributed) population because the maximum value in each case is at least several times the mean value and balancing values below the mean would be negative. Negative values are not allowed in such measurements. A plot of frequency curves for these data would show the mean to be larger than the mode or positive skewness.

III. EQUIPMENT AND PRINCIPLES

Analysis of air pollutants uses all the normal analytical principles, including spectral analysis, mass spectrometry, wet chemistry, chromatography, neutron activation, noise frequency and intensity, and radiation counting, plus the rather subjective olfaction.

The trend is toward highly sophisticated electronic analytical equipment. A well-equipped laboratory must have a number of pieces of equipment costing several thousands of dollars each. The role of the electronics technician is becoming increasingly important and the leasing of equipment, in which only operational time is paid for, is now popular.

A. Spectral Analysis

Each pure substance, element, or compound has a unique spectrum of electromagnetic energies which it can absorb or emit; therefore spectra are widely used for analysis. The energies contained in the electronic fields and the bonds are vibrational, rotational, or electrical. Each possible energy level has a unique quantum number. A substance can receive or emit only those quanta that take it to another level of excitement. Various energy sources are used to excite substances, and the amount of energy absorbed or emitted is measured by absorption or emission spectrometry. The amount of energy measured is related quantitatively to the substance identified by the spectrum. The principal limitation is that interferences from other substances must be removed.

The field of spectral analysis by photon absorption has made available elaborate instruments--spectrophotometers. The instrument may be limited to photons with wavelengths (energies) in one of the nominal electromagnetic bands, UV, visible, IR (near and far), and microwaves, or it may span two or more of these bands.

An absorption spectrophotometer consists of a source of the photons to be used, a prism or grating (sometimes both) or filter to obtain specific wavelengths within the band, and a device to measure energy passing through the sample, normally a bolometer. A bolometer absorbs photon energy in a temperature-sensitive resistor or thermistor which is part of a Wheatstone bridge. The bridge gives a measure of the energy received. Line spectra made by focusing transmitted light through a slit were formerly the principal output of spectrophotometers, but recorder traces of bolometer signals are now most frequently used. Line spectra are recorded on film, then analyzed by a densitometer (light transmission measurement).

Nearly all modern spectrophotometers scan automatically so that the full range of their wavelength capability is produced as a spectrum on a recorder chart. The wavelength is varied with time and changes along the recorder chart, the energy received being the signal tracked across the chart. Dual-beam operation is a popular mode and well worthwhile because it can remove the spectrum of the blank or carrier material. For dual-beam operation a splitter is used to send the signal through a blank half of the time and through the sample the remainder of the time. The separate signals are picked up by identical bolometers and the blank signal is removed such that the difference in the spectra is the signal recorded.

1. Ultraviolet

The usual UV-band instrument is a combination of UV and visible and goes from a wavelength of about 280 nm to the end

of the visible band at about 700 nm. The scale and the
descriptive literature show a range somewhat wider, to about
240 nm on the short end; however, the lamp is seldom capable
of providing sufficient energy for satisfactory operation at that
wavelength. The source of UV most used is the hydrogen lamp
(formerly it was the carbon arc); the visible source is a separate
lamp, and switchover from one to the other is accomplished
while scanning.

The UV spectrophotometer finds its widest application in
the measurement of polycyclic hydrocarbons--benzopyrenes,
anthracene, and so on. UV light is also used in fluorescence.
A fluorimeter detects visible photons coming from a material
that fluoresces when struck by UV light. This apparatus is
good for tracer work and has been used for dispersion studies.

2. Visible

Visible range spectrophotometers use a lamp (sometimes
two, one for blue and one for red) as a light source and photo-
cells as the measuring device. Full-range instruments use a
prism for specific wavelengths and have an adjustable-width
slit to select a portion of spectrum from the prism. The simplest
and cheapest instruments are filter photometers that use four or
five or more filters to obtain the desired wavelengths, have
adjustable lamp intensity, and a percent transmittance and/or
optical density meter. Filter photometers are priced at a few
hundred dollars, about 1/10 or less of the cost of the usual
scanning spectrophotometers.

The Beer-Lambert law was devised for colorimetric analysis such as that done with the filter photometer. The law is
the familiar first-order reaction equation that results from the
interactions that are random with respect to some parameter,
in this case with distance and concentration. The attenuation
of light is caused by interaction of the photon with particles along

the path. The chance for interaction is determined by the areal
density of particles in the light path through the sample, and
the decrease in light intensity is proportional to the light intensity.

$$-dI/d(cl) = kI, \tag{7-4}$$

where I = intensity of light, c = concentration of color, l =
length of path in sample, and k = constant of fit. The path
length for a given set of measurements is usually constant, and
the integration of Eq. 7-1 gives

$$I/I_0 = e^{-Kc}, \tag{7-5}$$

where I_0 = intensity of incident light and K = slope of plot of
I/I_0 (log) versus c.

EXAMPLE 7-1: If the extinction coefficient (K) is 2170 liters/mole
     for Fe-3 and an absorbance of 0.80 is read at 305 nm on
     a 20-ml sample containing the iron (converted to Fe-3)
     from a 100-m^3 sample, what is the ambient concentration
     of iron? This K is for base 10 logs.

$\log_{10} I_0/I = Kc$ $0.80 = 2170$ liters/mole X c
$c = 3.69$ X 10^{-4} moles/liter

Amount of iron:

3.69 X 10^{-4} moles/liter X 20/1000 liter X 56 gm/gm mole
$= 4.14$ X 10^{-4} gm

Concentration: 414 μg/100 m^3 = 4.14 μg/m^3

The meter of the filter photometer is usually calibrated to
read percent transmittance (0-100). It may also be marked to
show absorbance (optical density) (2-0), where absorbance
equals $\log_{10} I_0/I$, or with I_0 adjusted to 100% transmittance
through the blank, its maximum is 2.

Calibration curves for colorimetric analyses should give
straight-line plots of absorbance versus concentration on arithmetic graph paper, or percent transmittance versus concentration
on semilog paper. If commercially available semilog yields a

very flat slope over the full range of concentrations expected to
be found, and this is normally the case, graph paper should
be prepared to give a slope of near -1 (45°) by proportionally
changing the log scale. This technique gives graph paper that
may have only a fraction of 1 cycle on the log portion, for
example, from 70 to 100% or from 40 to 100%. The slope near
45° gives the minimum error in reading values from the graph.

The visible spectrum is especially good for analysis of
inorganics, but some organics give good color complexes also.

3. Infrared

A change in the dipole moment of an atomic system is
caused by or produces photons which are usually in the IR band.
For unsymmetrical molecules all normal vibrations are connected
to changes in the dipole moment, while symmetrical molecules
vibrate without dipole moment change; therefore unsymmetrical
molecules are IR-active and symmetrical molecules are not.
There are some notable exceptions to this generality, especially
carbon dioxide and water. Water is so active that samples are
often dried with silica gel, the air under the cover of the apparatus is dried, and samples are put in solvents other than water.
IR spectrometry is quite useful for identification and quantification of pure compounds. Catalogs of more than 20,000 spectra
are available (costing in excess of $2000). A spectrum may be
classified by number with each digit representing significant
identification information. The number can be used to narrow
the number of possible spectra down to a manageable size.

IR spectrometry is often used in conjunction with a gas
chromatograph. The pure material (plus carrier) is trapped at
the chromatograph exhaust or a rapid-scan IR spectrophotometer
is connected directly to the exhaust. The rapid-scan instrument
can give a qualitative spectrum in a few seconds. The spectrum

and the detention time give positive identification and the peak
height or area on the chromatograph chart indicates the quantity.

The IR spectrophotometer frequently changes from prism
to grating during the scan in order to obtain the best wavelength
resolution. Far IR is now being utilized in some analytical
instruments; it is often combined with near IR and perhaps more
of the spectrum.

Liquid and solid phase samples usually have short beam
paths, but the gas sample may have mirrors reflecting the beam
back and forth to obtain a length of many meters. A 10-m cell
is commercially available and is often used. Some installations
have 120- or 180-m paths. Gas may be put under pressures up
to a few atmospheres to give better sensitivity on the ambient
basis.

4. Microwave

Most microwave spectra come from changes in rotational
energy (moment of inertia) of gas molecules, in gases other
than diatomic ones because diatomic gases have no moment
of inertia about the internuclear axis. Microwave spectrometry
is used to identify gaseous materials and to detect free radicals.
It can handle small concentrations and samples but it is not
useful quantitatively. The principal applications are in physics,
not air pollution.

5. Other

The gamma spectrometer is used to obtain the energy
spectrum for wavelength photons shorter than the UV. It is a
pulse height analyzer that measures the energy from absorption
of the photon in a scintillator or in a solid state detector.

The atomic absorption spectrophotometer differs from the
instruments described above in that it has a lamp made with the
metal being run; therefore the spectrum coming through the sample

is that for the sample itself. This unit uses a flame to burn the sample; the source light is passed through the flame; and the absorption of a characteristic peak is checked. The method works best for copper, iron, arsenic, lead, and other such metals.

The flame spectrometer burns a sample in a flame, and the light from the combustion is analyzed for the spectrum of the material being tested. This method is similar to the atomic absorption technique and works best for the metals, including sodium, potassium, magnesium, and so on.

B. Mass Spectrometry

In mass spectrometry the molecules (atoms) of the sample substance are ionized by electron bombardment and then the masses of the ions determined. The latter can be accomplished by various means, the most popular being the use of a magnetic field to bend the path of the ions, which are moving with a certain velocity, into the collector electrode.

$$R = \frac{mv}{He} = \frac{1}{H}\sqrt{2V\frac{m}{e}}, \qquad (7\text{-}6)$$

where R = radius of curvature, m = mass of ion, v = velocity of ion, H = magnetic field strength, e = charge of ion, usually 1, and V = potential applied to accelerate ion. The ion current from the collector electrode gives a measure of the number of the ions with a mass which made the proper curvature for collection. Other methods of mass determination include the measurement of the velocity (time of flight, $t = k\sqrt{m/e}$), or separation of the ions by a radiofrequency ac field or by magnetic bending through a spiral while accelerating (as in a cyclotron) and splitting the particles of desired mass. A good instrument can resolve peaks about 0.1 amu apart. Some mass spectrometers selling for less than $10,000 have recently become available; however, the price may range up to several times this amount.

The mass spectrometer can be used to obtain the mass of sample material, then fragments of the material. By serially reducing the sample material into smaller parts, it can often be identified by the mass spectrometer; however, its most effective use probably lies in its ability to identify the isotopes present in a known compound or to follow a gas chromatograph that separates a known gas.

C. Chromatography

Chromatographic methods make use of the fact that different molecules migrate at different rates when carried along by a fluid through some medium. The molecules seem to adhere momentarily and then move on. After moving for some distance, the material separates into the various components. The pure compound is mixed with the carrier gas but is not mixed with any interfering substance. There are several types of chromatography in common usage, including paper, column, gas, gas-liquid, and thin-layer.

1. Paper Chromatography

Paper chromatography is done with an absorbing paper, such as filter paper, and a liquid solvent. The exact method varies widely. It is difficult to obtain quantitative results and to do so requires good, careful techniques. A strip of paper may be suspended, a portion of the sample material put on the paper near its top, and solvent allowed to flow down over the spot. The paper may be suspended with its end in the solvent so that capillarity moves the material upward. Cabinets are available to hang such papers in. After migration has been accomplished, the location of the material of interest must be found and measurement of the amount of material made.

Detection methods in the past chiefly involved color development, with estimates of the size and intensity of color spot being the quantitative measure. Now radioactive labeling is often done, permitting a count to be made of the spot and giving much better quantification of the material.

The Weis ring oven is used to speed up paper chromatography. The sample is placed in the center of a piece of filter paper, solvent is flowed over the spot, and the material is heat dried as it moves outward on the paper.

2. Column Chromatography

A sample may be washed through a column packed with fine granular material coated with an adsorbent material and emerge in its separate fractions at the bottom of the column. The portion with the proper residence time is caught and analyzed for the amount of the material of interest.

3. Thin-Layer Chromatography

Thin-layer chromatography (TLC) is carried out by spreading a thin coating of material on glass and allowing the sample and solvent to migrate in the coating. TLC has become quite popular in the last several years, especially in biochemistry work.

4. Gas Chromatography

Gas chromatography (GC) is the most versatile and widely used method of chromatography. The gas chromatograph has an inert carrier gas, normally helium, flowing through identical, packed columns which are made by filling 1/4 in. tubing about 4 ft in length; the sample is injected into one of the columns, and the difference between the gases emerging from the columns is detected in a manner that gives an electrical signal which is recorded. As each component of the sample comes through, a peak is registered on the recorder chart; the height of the peak or the area under the peak is used to quantify the component and its residence time in the column, injection to emergence, for tentative identification of the substance. If no other material

with a residence time similar to that of the material being analyzed
is present, this identification may be all that is necessary.

Column packings are quite varied, usually consisting of an
inert matrix such as crushed brick coated with a chemical such
as silicone rubber. Very long capillary tubes are sometimes
used instead of packed columns, especially for materials that
cannot be separated very well with the usual columns. An
example of such a material is an amine odor.

The sample injected may be either a gas (milliliter amounts)
or liquid (microliter amounts). The temperature of the injection
port vaporizes the sample. The columns are contained in an oven
and the injection port is kept hot. Temperature programming may
be used to drive off fractions with higher and higher boiling
points. The temperature control on a research-type gas chromatograph gives considerable versatility to a single type of column.

The detectors normally used are thermal conductivity,
flame ionization, and electron capture. Thermal conductivity
(with two thermistors in a bridge circuit, one in the carrier gas
and one in the carrier plus sample) was the early method and is
quite satisfactory for some materials. Sensitivities can be
lowered by a factor of $10^3 - 10^6$ through the use of flame ionization or electron capture instead of thermal conductivity. In
flame ionization gas coming from the columns is split and portions are passed through hydrogen-oxygen flames to ionize the
components; then an electrometer is used to measure the ion
current. Electron capture uses an ionizing radiation source
(^3H or ^{90}Sr) to ionize the molecules which are then measured
by the electrometer.

Many accessories, columns, and techniques have been
developed for the gas chromatograph. In addition to the applications of the gas chromatograph alone, there are many uses for it

in combination with the IR spectrometer and other instruments.
The analyses of hydrocarbons, chlorinated hydrocarbons, pesticides, alcohols, esters, ketones, and many other classes of
substances may be either carried out completely or materially
aided by the use of GC.

D. Wet Chemistry

Materials removed from the atmosphere by sampling are
often analyzed by wet chemistry methods. The materials may
be absorbed in liquid for sampling, or if the materials are removed
in another form, they are frequently put into solution for analysis.

The equipment for wet chemistry is generally the same as
for all chemistry laboratories. It should include laboratory glassware (pipets, beakers, graduated cylinders, flasks, evaporating
dishes, funnels, stills, reflux condensers, test tubes, and so
on), analytical balances, a supply of chemicals, ovens, a
desiccator, a muffle furnace, pH meters, and a filter photometer
or visible spectrophotometer. A considerable amount of equipment is available for automated wet analyses as is noted in
Section IV, D.

Color formation is the most frequently used type of chemical
reaction for analysis. Such analyses give a color with a concentration that is a direct function of the amount of test material
present. The color is compared with a calibration standard to
obtain the concentration of the pollutant. The concentration is
easily quantified by a filter photometer or visible spectrometer.
An example of a colorimetric analysis is the dithizone test for
lead (see Chapter 9).

Neutralization reactions are sometimes used in air pollution analysis, but the results are usually nonspecific. Exceptions are such tests around an acid or alkali plant, where the
material is known and only the amount present is in question.

Precipitation reactions are not used very much in air pollution analyses because an easier method is usually available. In precipitation reactions an insoluble salt is formed from the material (ion) being tested. The precipitate must be filtered, washed, dried, and weighed.

Indirect chemical reactions are exemplified by the potassium iodide test for ozone.

$$O_3 + 2KI + H_2O \rightleftharpoons O_2 + 2KOH + I_2 \qquad (7\text{-}7)$$

The iodine in the resulting solution is analyzed as a determination of the amount of ozone, although ozone is not in the product of analysis.

E. Electrochemistry

Electrochemical methods measure the mobilities and activities of ions in solutions (electrolytes) by measurements of potentials and currents. The gross effect of all the ions present may be measured by the resistance (conductance) between two inert electrodes. Specific ion measurements use active electrodes in either one or both of the positions and often are covered with a membrane which keeps out interfering substances. The active electrode has the elemental material in chemical equilibrium with ions of the same material in solution.

The Nernst equation is used to describe the potential between an elemental electrode and its ion in solution (a half-cell)--the potential to transfer electrons from the oxidant to the reductant.

$$E = E_0 + \frac{RT}{zF} \ln \frac{[I_o]}{[I_r]}, \qquad (7\text{-}8)$$

where E = electrode potential (V), E_0 = reference electrode potential (V), (see Table 7-2), R = universal gas constant = 8.316 joules/gm mol-$^\circ$K, T = absolute temperature ($^\circ$K), z = valency of ions, F = Faraday constant = 96,500 coulombs, and

TABLE 7-2

Electrode Potentials of Some Metals

Electrode	E_0 (V)	Reaction
K	-2.92	$K \longrightarrow K^+ + \epsilon$
Ca	-2.87	$Ca \longrightarrow Ca^{2+} + 2\epsilon$
Na	-2.71	$Na \longrightarrow Na^+ + \epsilon$
Zn	-0.76	$Zn \longrightarrow Zn^{2+} + 2\epsilon$
Fe	-0.44	$Fe \longrightarrow Fe^{2+} + 2\epsilon$
Cd	-0.40	$Cd \longrightarrow Cd^{2+} + 2\epsilon$
Ni	-0.25	$Ni \longrightarrow Ni^{2+} + 2\epsilon$
Sn	-0.13	$Sn \longrightarrow Sn^{2+} + 2\epsilon$
Pb	-0.12	$Pb \longrightarrow Pb^{2+} + 2\epsilon$
H_2	0.00	$1/2\ H_2 \longrightarrow H^+ + \epsilon$
Cu	0.34	$Cu \longrightarrow Cu^{2+} + 2\epsilon$
Hg_2	0.79	$1/2\ Hg_2 \longrightarrow Hg^{2+} + \epsilon$
Ag	0.80	$Ag \longrightarrow Ag^+ + \epsilon$

$[I_r]$ and $[I_o]$ = concentrations of the reduced and oxidized forms of the ion in solution. For values of E_0, see Table 7-2.

EXAMPLE 7-2: What potential exists between the glass electrode and the saturated calomel electrode of a pH meter when the pH is 4?

$$E = E_0 + 2.303\ (8.316)(293)/[1\ (96,514)]\ \log_{10} Q$$
$$= E_0 + 0.058\ \log_{10} Q$$

The pH meter setup is as follows:

$Hg-Hg_2Cl_2$--Sat KCl/sol'n of test/glass--0.1 M HCl--
                                                    Ag-AgCl

$$E = E_{cal} + E_{jct} - E_g - E_{sil} = E_{char} - E_g$$

The characteristic potential (E_{char}) for the cell is on the order of a few hundred millivolts. E_{cal} = 246 mV and

$$E_g = E_0 + 0.058\ \log_{10} [H^+] \text{ and for glass, } E_0 = 0$$
$$= -0.058 \times pH = -0.058\ (4) = -232\ mV$$

Existing potential for the system is

$E = E_{char} + 232$ mV.

The galvanic cell is made up of two half-cells and the potential is the difference between the two Nernst potentials.

In polarography a small potential is impressed across a polarizable electrode and a dropping mercury electrode (nonpolarizable). The current flow is measured with a reflected light beam or other galvanometer.

F. Odors

Odor measurements may be described in two groups, namely, those made with the nose (olfaction) and those made with analytical instrumentation. The most important equipment for all olfaction measurements is that which provides (1) odor-free space for performing the tests and (2) sniffs of the odor (measured sniffs for odor unit determinations). The odor-free space may be a room, but more often it is smaller; it may be a hood or even a face mask. Chemical analysis of odors often requires the most sensitive analytical methods available and even then there is difficulty. This is a direct consequence of the extreme sensitivity of the human sense of smell which may be stimulated by several molecules of an odor, versus the usual picogram or larger quantities detectable by instrumental analysis. Both measurements have their faults and there are those who champion one or the other or a combination of the two.

For subjective measurements to be meaningful, there must be definitions of quality and quantity, that is, what is it and how much there is. Unfortunately, many systems have been devised and they often contain three or even four parameters for characterizing odor. For relevancy, chemical analysis of the components should relate to the odor sensation. This might not be too difficult in the case of some single odors; however, the sensations of multiple odors (masking, counteraction,

potentiation, and so on) and fatigue from long exposure or overwhelming concentration seem impossible to link to analyses.

1. Quantity

A starting point in relation to quantification of odor seems
to be the definition of an odor threshold--there are two: one for
perception and one for recognition. Furthermore, the sense of
smell results must be a statistical average because of biological
variability. The thresholds normally used are those for 50% of
the population and are sometimes abbreviated as PPT or PPT_{50}
for the population perception threshold and PIT or PIT_{50} for the
population identification threshold. When the word threshold
is given without qualification, it is usually the recognition
threshold. The recent Manufacturing Chemists' Association
(MCA) study (G. Leonardos and associates) determined the PIT_{100}
for 53 chemicals as shown in Table 7-3. Because the numbers
are for recognition and for 100% of a test population, they should
be rather large. A check of about 10 of these values with former
listings shows no consistent pattern of differences except that
when the MCA values are larger they are larger by factors of the
order of three, and when they are smaller they may be smaller
by factors of hundreds.

A number description of the quantity of odor has long been
used. In it numbers describe the odor intensity according to the
following. 0, no odor; 1, barely detectable odor; 2, definite
odor; 3, strong odor; and 4, overpowering odor. Some investigators state that 1 corresponds to the PIT_{50} and 1/2 to the PPT_{50}.
Other workers object to only five classes, but that should present
no problem as they can use fractions for the intermediate classifications. According to the Weber-Fechner law, the odor sensation should be proportional to the logarithm of the odor concentration; that is, the odor intensity numbers on an arithmetic scale
should plot a straight line versus the odor concentrations on a

log scale. One of the major drawbacks in quantification by smell is the subjective reaction--an observer may be too eager to classify an objectionable odor as overpowering, and the reverse may occur for pleasant odors.

Another approach toward quantification uses odor units. An odor unit is defined as the amount of odoriferous material(s) present in 1 ft^3 of air at the PPT_{50}. The use of the volumetric term in the definition permits the presentation of the amount of odorous emissions through the use of a single number--the odor units per minute = (o.u./scf) X scfm.

Odorous materials analyzed by methods other than olfaction should be determined in parts per million (volume) or micrograms per cubic foot as is done for other air pollutants.

2. Quality

Many descriptive word systems have been suggested to describe the nature of an odor. Sometimes the description depends solely upon the adjective vocabulary of the observer, but in most cases there are some recurring terms. Table 7-3 uses a rather liberal sprinkling of the terminology.

A quite logical description of odor quality was put forth by Crocker and Henderson. It uses four words (fragrant, acid, burnt, and caprylic) for component types, and the numbers 0 to 8 to rate the relative importance of each component to an overall odor. Acetic acid was described as 3803 and methyl salicylate as 8453. The system has been in existence for a long time and it is still not widely used.

Chemical analysis is usually aimed at the generic identification of an odorous material. As mentioned previously, it is difficult to describe an odor sensation from the generic names except for odors produced by only one substance.

TABLE 7-3

Odor Thresholds for Pure Chemicals[a]

Chemical	Odor threshold (ppm$_v$)	Odor description[b]
Acetaldehyde	0.21	Green sweet
Acetic acid	1.0	Sour
Acetone	100	Chemical sweet, pungent
Acrolein	0.21	Burnt sweet, pungent
Acrylonitrile	21.4	Onion-garlic pungency
Allyl chloride	0.47	Garlic-onion pungency, green
Amine, dimethyl	0.047	Fishy
Amine, monomethyl	0.021	Fishy, pungent
Amine, trimethyl	0.00021	Fishy, pungent
Ammonia	46.8	Pungent
Aniline	1.0	Pungent
Benzene	4.68	Solvent
Benzyl chloride	0.047	Solvent
Benzyl sulfide	0.0021	Sulfidy
Bromine	0.047	Bleach, pungent
Butyric acid	0.001	Sour
Carbon disulfide	0.21	Vegetable sulfide
Carbon tetrachloride (chlorination of CS_2)	21.4	Sweet, pungent
Carbon tetrachloride (chlorination of CH_4)	100	--
Chloral	0.047	Sweet
Chlorine	0.314	Bleach, pungent
Dimethylacetamide	46.8	Amine, burnt, oily
Dimethylformamide	100	Fishy, pungent
Dimethyl sulfide	0.001	Vegetable sulfide
Diphenyl ether (perfume grade)	0.1	--
Diphenyl sulfide	0.0047	Burnt rubbery
Ethanol (synthetic)	10	Sweet

Ethyl acrylate	0.00047	Hot plastic, earthy
Ethyl mercaptan	0.001	Earthy, sulfidy
Formaldehyde	1.0	Hay/strawlike, pungent
Hydrochloric acid gas	10	Pungent
Hydrogen sulfide (from Na_2S)	0.0047	Eggy sulfide
Hydrogen sulfide gas	0.00047	--
Methanol	100	Sweet
Methyl chloride	>10	--
Methylene chloride	214	--
Methyl ethyl ketone	10	Sweet
Methyl isobutyl ketone	0.47	Sweet
Methyl mercaptan	0.0021	Sulfidy, pungent
Methyl methacrylate	0.21	Pungent, sulfidy
Monochlorobenzene	0.21	Chlorinated, moth balls
Nitrobenzene	0.0047	Shoe polish, pungent
Paracresol	0.001	Tarlike, pungent
Paraxylene	0.47	Sweet
Perchloroethylene	4.68	Chlorinated solvent
Phenol	0.047	Medicinal
Phosgene	1.0	Haylike
Phosphine	0.021	Oniony, mustard
Pyridine	0.021	Burnt, pungent, diamine
Styrine (inhibited)	0.1	Solventy, rubbery
Styrene (uninhibited)	0.047	Solventy, rubbery, plasticy
Sulfur dichloride	0.001	Sulfidy
Sulfur dioxide	0.47	--
Toluene (from coke)	4.68	Floral, pungent, solventy
Toluene (from petroleum)	2.14	Moth balls, rubbery
Toluene diisocyanate	2.14	Medicated bandage, pungent
Trichloroethylene	21.4	Solventy

[a]From G. Leonardos et al., "Odor Threshold Determinations of 53 Odorant Chemicals," <u>J. Air Pollution Control Assoc.</u>, 19:2, 91-95, February 1969.

[b]Other than chemical name.

G. Radiation

The physical effects produced by radiation interactions with matter which may be used in its measurement are the heat rise (bolometry or calorimetry), ionization (counts or currents), storage followed by release of energy (thermoluminescence), electrophotometer conversion of photons to an electrical signal (light meter), crystalline color change (salt or glass), and radiation pressure (radiometry). Radiation interactions also cause several chemical effects which may be used to measure radiation, including oxidation (Fe^{2+} to Fe^{3+}) and reduction (film emulsion), polymerization (gelation or viscosity increase) and chain breakage (viscosity decrease), and cross-linkage (stiffening of fiber) and side-group scission (increasing flexibility of fiber). More energetic radiation is usually easier to measure than weak radiation. Ionizing radiations, corpuscular and electromagnetic, can be measured by a wide range of equipment and techniques.

A very large number of measurement units have been defined for radiation. A few of them are presented here as succinctly as possible. There are measures of radiation flux (number per square centimeter per second) without specific inclusion of energy. It is relatively easy to count ionizing radiations, and the flux can often be related back to a source strength. The activity of the source is given in curies (Ci) where 1 Ci equals 3.7×10^{10} disintegrations per second. The visible light flux is defined in candlepower; the flux of 1 candle is equal to the average emitted by 45 standard sources at the National Bureau of Standards. Derived units that relate to the candle are the lumen (flux in 1 steradian solid angle from 1 candle) and the unit for illumination, the footcandle (illumination of 1 lumen/ft^2).

Most other radiation fluxes are given in energy per unit area, such as watts per square centimeter per second used for

microwave, UV, and IR; calories per square centimeter per
second for laser radiation.

Fluxes can be converted to doses in many cases simply
by multiplying by the time of exposure; however, for ionizing
radiation additional units have been defined for the exposure dose
and the absorbed dose, the roentgen and the rad, respectively.
The roentgen is that amount of x or γ radiation that gives 1 esu
of charge to 1 cm^3 of STP air. The rad is the radiation absorbed
dose of 100 ergs/gm of absorber.

The common equipment for measuring radiations includes
the bolometer, Geiger, proportional, and scintillation counters,
ion chambers and electrometers, light meters, phosphate glass,
film, and chemical solutions. The measurement of radiations
is not important enough to warrant description of this equipment
here. The pulse-height analyzer has been developed largely
for ionizing radiations but has wide application in particle size
counting, time-of-flight studies, and other measurements that
give an electrical pulse.

H. Noise

Noise is a subjective response as is odor; however, noise
measurement has progressed far beyond that available for odors.
The vibrations of air molecules produce waves of pressure change
that make up noise. The pressure changes are picked up by a
transducer and converted to an electrical signal in a microphone.
The electrical signal is then amplified and measured. Both
components of noise, frequency and intensity, can be measured
with an accuracy that suffices for air pollution purposes.

Noise vocabulary has a large number of specialized words
in addition to the technical terms applied in other fields as well.
The frequency of noise is now given in hertz (Hz) to replace the
previously used and perfectly descriptive term cycles per second
(cps). Decibel is another term used in fields other than noise

measurement. The decibel is particularly well suited to measurements over a wide range of values. Our hearing range is very
wide (approximately a ratio of 10^{20}) and follows the Weber-
Fechner law discussed above in relation to odors.

The decibel is a unit for relating relative intensity levels.

$$dB = 10 \log_{10} \frac{I}{I_0}, \qquad\qquad (7\text{-}9)$$

where dB = decibels of intensity (or gain), I = intensity measured,
and I_0 = reference intensity. The lowest audible sound should
serve as the reference. Noise intensity levels may be based on
power levels (L_w, formerly PWL) or pressure levels (L_p, formerly
SPL). The reference intensity for L_w (0 dB) is usually taken as
0.1 pW (10^{-13} W) per square foot. Note that the intensity is
power per unit area, whereas source strength is power.

$$L_w = 10 \log_{10} W/W_0, \qquad\qquad (7\text{-}10)$$

where L_w = sound intensity, power level basis (dB), W = power
of measured sound (W/ft^2), and W_0 = reference power level =
10^{-13} W/ft^2. The pressure intensity level is normally referenced
to an rms pressure of 0.0002 μbar, and since μW/ft^2 = (1.52 μbar)^2,
the equation for sound intensity based on pressure is

$$L_p = 10 \log_{10} (p/p_0)^2 = 20 \log_{10} p/p_0, \qquad\qquad (7\text{-}11)$$

where L_p = sound intensity, pressure basis (dB), p = rms pressure
of measured sound (μbars), and p_0 = reference rms pressure.

For a given noise the dB values for L_w and L_p should be
equal because the defining equations take account of the relation between power and pressure and both equations use the
threshold of hearing as the reference intensity.

Octave band analysis has been widely used in past noise
analyses. The American standard bands are divided by frequencies
of 75-150, 150-300, 300-600, 600-1200, 1200-2400, 2400-4800,
and 4800-9600 hz. Current usage is the "center-line" frequency

for the band; these are 125, 250, 500, 1000, 2000, 4000, and 8000. The American standards also set up three weighting response curves (see Fig. 7-2a). An octave band analysis may be converted to equivalent A band by using Fig. 7-2b.

There are three types of equipment most often used in noise measurements--a survey meter, an impact device, and an octave band analyzer. These instruments are discussed briefly in techniques of noise measurement (Section IV, J).

IV. TECHNIQUES OF ANALYSIS

This section deals with applications of the principles and equipment described above. It is as brief as possible in covering material considered needed for general guidance in measurements and for specific help in running the specific tests of Chapter 9.

A. Titrations

Chemical analyses resulting in neutralization or the formation, release, or uptake of a substance that can be detected by indicators or by electrochemistry are often carried out by titration. The amount of titrant used to reach the end point is the measure of the material that was present. The method is used to standardize acids, bases, and analytical chemicals.

The reacting equivalents in titrations are determined from

$$N_1 V_1 = N_2 V_2, \tag{7-12}$$

where N = normality (gram equivalent weights per liter) of solution, and $N_i V_i$ = number of equivalents. In Eq. 7-6 the volumes may be in liters and $N_i V_i$ in gram equivalent weights, or the volumes may be milliliters and $N_i V_i$ in milliequivalents.

A few numbers that will bear mentioning are some common valences and normalities. H, Na, K, Cl, and NO_3 all have valences of 1; and Ca, Mg, O, CO_3, and SO_4 all have valences of 2. Concentrated sulfuric acid has a normality of about 36 and concentrated hydrochloric acid is about 14 N.

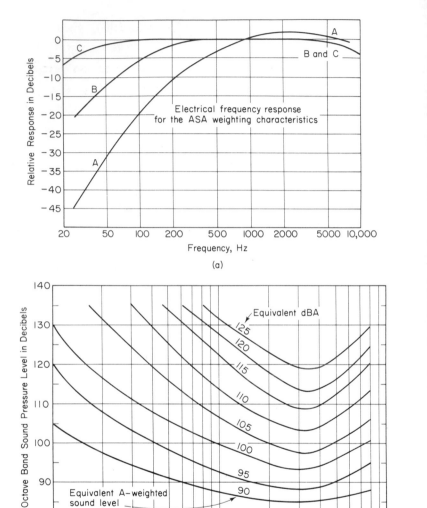

FIG. 7-2. Noise weighting curves. (a) American Standards
Association sound level meter requirements. From A. P. G.
Peterson and E. E. Gross, Jr., Handbook of Noise Measurement,
5th ed., General Radio Co., West Concord, Massachusetts, 1963
(b) A-Weighted sound (for Walsh-Healy compliance). From "The
Walsh-Healey Public Contracts Act," Noise News Rev., 1:1, 6,
February 1970.

<u>EXAMPLE 7-3</u>: How much water should be added to 400 ml of an iodine solution to bring its normality to 0.05 if 5.00 ml of the solution required 16.40 ml of 0.025 N sodium thiosulfate to remove the color of the starch indicator?

$$N_1V_1 = N_2V_2 \quad N_1 \ (5.00 \ ml) = 0.025 \ (16.40 \ ml)$$
$$N_1 = 0.082$$

$$0.05 \ (V_1) = 0.082 \ (400 \ ml)$$
$$V_1 = 656 \ ml$$

Volume to be added: 656 ml - 400 ml = 256 ml

Titrations are carried out by releasing the titrant from a buret which is carefully read before and after the titration to 0.01 ml. The curve of the meniscus is read with the eye on a level with the liquid. The titration for a color change end point is carried out against a white background; a large porcelain evaporating dish is an excellent titration container. The additions should be dropwise near the end point. If unsure whether or not the end point has been reached, read the buret, then add more titrant and watch for further change.

Indicators are described as they are needed in specific tests. Some common pH indicators and their ranges of usefulness are: thymol blue-1 (1.2-2.8, red-yellow); methyl orange (3.0-4.4, red-yellow); methyl red (4.4-6.2, red-yellow); bromthymol blue (6.0-7.6, yellow-blue); phenol red (6.8-8.4); phenolphthalein (8.2-10.0, colorless-red); trinitrobenzene (12.0-14-0, colorless-orange). End points may be indicated by pH meters or other electrochemical measurements.

B. Electrochemistry

Air pollutants put into aqueous solution or suspension may be identified, quantified, or characterized as to some property by electrochemistry. The measurements made are conductivity, ion mobility, electrophoretic velocity of particles, zeta potential of colloids, streaming current counterions (colloids immobilized with fluid passing, pick up counterions and cause current), or ion activity by specific ion electrodes.

Conductivity measurements are used in most ambient sulfur dioxide monitoring. The sulfur dioxide is dissolved in a solution whose conductivity is measured in two types of instruments. The Thomas autometer type absorbs the sulfur dioxide in a conductivity cell which is periodically flushed with clean acid-peroxide. The Titrilog type instruments operate somewhat similarly except that the absorption of the sulfur dioxide upsets a potential balance on the cell and generates a halogen to restore the balance.

Specific ion electrodes (SIE) are rapidly being developed and are now available for several cations and anions. SIE measure the ionic activity or tendency of the ions to react; the activity is practically equal to the ionic concentration for dilute solutions (to about 0.01 M) and is functionally related to higher ionic concentrations, making calibration of electrode potential (millivolts) versus ion concentration (moles per liter) possible.

The specificity of a glass electrode for H^+, Na^+, K^+, and Li^+ is controlled by the relative amounts of Na_2O and Al_2O_3 in the glass. Liquid-liquid membrane electrodes are filled with a liquid ion-exchange material which is specific for Cu^{2+}, Mg^{2+}, and Ca^{2+}. Membranes of silicone rubber or single crystal solid state are impregnated with specific precipitates for such ions as F^-, Cl^-, Br^-, I^-, and S^{2-}. Electrodes can be made for NO_3^-, SO_4^{2-}, ClO_4^-, enzymes, and antibiotics, and an almost endless variety seems to be possible in the future. Measurements of dissolved oxygen have been made by this method for several years.

The most often used potentiometry is in the measurement of pH. In theory the pH meter measures H^+ concentration by using a bubbling hydrogen electrode with a calomel reference electrode, but in practice a glass electrode gives the same results as the bubbling hydrogen and is of much simpler construction. E_0 for the saturated calomel electrode is 0.246 V at 25°C.

The pH meter circuitry can be used for other potentiometric measurements by changing to the proper electrodes. For example, the substitution of electrodes is often carried out to measure the oxidation-reduction potential of waste water.

C. Color Analysis

Many quantitative chemical analyses are carried out by complex formation to yield a color with an intensity proportional to the concentration of reacting material present. An excess of the complexing reagent is normally used to assure complete reaction.

One of the best complexing agents is dithizone (diphenyl-carbazone). It is used to form color complexes with the heavy metals. By adjustment of the pH, altering the valences of interfering metals, or masking extraneous reactions, dithizone can be used for lead, cadmium, manganese, tin, zinc, bismuth, thallium, cobalt, iron, copper, silver, mercury, gold, nickel, indium, and platinum.

The tests for sulfur dioxide, nitrogen dioxide, and aldehydes are color-complexing reactions, as are other standard tests. Indicator tubes are color change devices that are made to be specific for the gas tested. Mine Safety Appliances and Unico (Kitagawa) produce tubes available for measuring at least 40 different chemicals. One of their principal drawbacks is lack of sensitivity. They are made to measure industrial hygiene levels, not ambient air levels.

After development of a color in an analysis, its intensity is measured by a spectrophotometer, a colorimeter (filter photometer), or visual comparison. For instrumental methods a calibration curve is prepared by mixing known amounts of the test material, developing the color, running optical densities or light transmittances, and plotting a calibration curve on semilog paper as described previously. Occasionally, instead of running known materials through the analytical procedure for the calibration, known dilutions of some colored material are used. This is especially useful for visual analysis in which the true color may fade quickly. Visual comparison methods often require long light paths for obtaining good results. Nessler tubes give a light path of 20 cm or more. They should be viewed against a white background.

Colorimetry measurements should be made with a matched set of cuvets. Cuvets are so expensive in matched sets and are so often broken that the tendency to substitute other glassware, often test tubes, is overwhelming. Some rather low-cost tubes will often suffice if the analyst is careful to check them by rotation in the holder to determine the variation in the glass itself and check them one against another for matching, or use the same tube for all of the measurements.

D. Automated Measurements

Air pollution analyses may be carried out automatically and the results recorded and/or telemetered to a data station. The field is not a new one, but nearly all of the significant developments have been made in the last several years. CAMP, which was instituted by the Public Health Service to gather air pollution data on the major pollutants in our principal cities, was responsible for much of the early development of automated air pollutant analyses. The aim of CAMP was to measure carbon dioxide, sulfur dioxide, nitric oxide, nitrogen dioxide, ozone, total oxidants, and total hydrocarbons with a portable unmanned station which was to be checked once daily by a technician.

There were enough problems to warrant manning the stations
full time, but the stations succeeded in obtaining more continuous and more reliable data than had been available.

Automatic instrumentation can measure conductivity (Thomas
autometer type of sulfur dioxide instrument), potential (titrilog
type of sulfur dioxide instrument), color (Technicon Autoanalyzer),
amperometric titration (Mast ozone), nondispersive spectrophotometry (carbon monoxide by IR), suspended particulates
and soiling index (paper tape with auto read of light transmittance
or reflectance and β radiation transmission), plume opacity
(transmissometer), and radioactivity (auto air monitors). All
of the foregoing instrumentation are available, along with other
unnamed equipment. Automated equipment can be made for
almost any application. Remote sensing of the environment by
satellite-carried analyzers has been done and is being planned
on an increasing scale.

Automated instruments can have a built-in calibration
check for proper operation. They can run continuously or intermittently and they may save money in some applications. Continuous measurements give a valuable record to an industry for
use in monitoring its pollutant emissions, a record which is
useful for regulations compliance and for a check on process
control.

E. Gravimetry

In addition to the weighing of precipitate mentioned in
Section II, D, there are other gravimetric determinations often
made in air pollution analyses.

1. Gravimetric Measurements

Chief among the gravimetric tests are the weight of dustfall and of suspended particulates and its fractions. Organic
(benzene-soluble), inorganic (ash after burning), and other

fractions may be determined gravimetrically. Concentrations
of the gaseous constituents is often given in micrograms per cubic
meter, although the only weighing is done in making up the
stock solution or standard. Further discussion of commonly
used gravimetric analyses appears in Chapter 9.

2. Equilibrated Weighing

The receptacle for the material to be weighed, whether
it collects the material (as a filter) or receives the material by
transfer (as an evaporating dish), is carefully tared beforehand.
The clean receptacle is placed in a $103^{O}C$ oven for sufficient time
to reach $103^{O}C$ throughout and remain there for 30 min. The
receptacle is removed from the drying oven and placed in a
desiccator to cool. When the temperature of the receptacle
is near room temperature, it is removed from the desiccator and
weighed on an analytical balance. Operational instructions for
the balance should be scrupulously followed. After the sample
is on or in the tared receptacle, the drying, cooling, and weighing procedures listed above are repeated. The gain in weight is
the weight of the dry sample.

F. Activation Analysis

Bombardment of atomic nuclei with particulate or energetic
electromagnetic radiations can cause nuclear reactions which
make the absorbing nuclei radioactive. This procedure is often
the most sensitive analytical method available. The activations
are nearly always carried out by neutrons in a reactor, or a
neutron source, or by charged particles in an accelerator. This
discussion centers on thermal neutron activation in a nuclear
reactor, but the same approach may be used for other activations.

Many materials have sufficient cross sections (probabilities for neutron-gamma reactions (n, γ) to make neutron activation
a good analytical method. The sample is put into a flux of ther-

mal neutrons (in the core of a reactor), and all of the substances in the sample become radioactive to varying degrees. By measuring the energy spectrum of the γ rays emitted from the activated material with pulse-height analysis, it is often possible to identify and quantify the various elemental substances present in the sample.

The amount of activity induced in a given element is calculable from

$$A = N\phi\sigma (1 - e^{-0.693\, t/T_r}), \tag{7-13}$$

where A = activity of isotope of interest (dps), N = total number of atoms of isotope precursor present, ϕ = thermal neutron flux (neutrons/cm^2-sec), σ = cross section for activation to isotope of interest (barns; cm^{-2}; 1 barn = 10^{-24} cm^2), t = time that activation has progressed (sec), and T_r = radioactive half-life of isotope of interest (sec).

EXAMPLE 7-4: What is the activity as ^{65}Zn (T_r = 245 d) in 1 gm of stable zinc if ^{64}Zn has an n, γ cross section of 0.47 barns and the reactor flux is 10^{13} during an irradiation of 20 days?

$$A = N\phi\sigma (1 - \exp -0.693\, t/T_r) \quad N = (6.02 \times 10^{23}/65.4)$$
$$(48.9\%)$$
$$= 4.50 \times 10^{21} \text{ of } ^{64}Zn$$
$$(48.9\% \text{ abundance})$$

$$A = 4.5 \times 10^{21} \times 10^{13} \times 0.47 \times 10^{-24}$$
$$(1 - \exp -0.693\, (20)/245) = 2.12 \times 10^{10} (1 - 0.945)$$
$$= 1.16 \times 10^9 \text{ dps} = 0.314 \text{ curies} = 314 \text{ mCi}$$

It is seen in Eq. 7-7 that irradiation beyond 5 or 6 half-lives does not further increase the activity. Decay after removal from the reactor must also be calculated by multiplying A by the exponential factor, with t being the time since activation ceased.

The sensitivity of neutron activation analysis depends upon the cross section, the available flux, and the efficiency

of pulse-height analysis. The best sensitivities are for arsenic,
gold, silver, indium, rhodium, iridium, and dysprosium.

The area of activation analysis was vastly enhanced by
the development of solid-state detectors for pulse-height
analysis. Although the efficiencies of detection are not generally
as good as was possible for other detectors, the resolution of
the energy peaks was improved by a factor of nearly 100 (about
8-0.1%). This means there is less overlapping of the numerous
peaks usually present and therefore interference in the analysis.
Prior to the use of solid-state detectors, chemical separations
were often required before practical pulse-height analysis could
be performed.

G. Radioactive Tracer Analyses

Radioactive tracers may be used to advantage in some
dispersion experiments and in radiochemical analysis procedures.
Probably the most valuable information gained by radioactive
tracer of dispersion has been from the long-term fallout of radioactive particles injected into the stratosphere by atomic tests.
Dispersion coefficients were calculated for the Oak Ridge stack
discharges. For tracer applications in the laboratory, the yield
of chemical reactions can be checked by adding a known amount
of tagged material that has the same chemical form as the substance being tested. The yield of the radioactive substance is
easily checked, and the yield of the stable material is assumed
to be the same.

H. Odors

Odor units in an air sample are measured by diluting a
volume of the odorous air with odor-free air until the odor
threshold is reached. The odor units are simply the total volume
of the threshold dilution divided by the volume of sample in the
dilution. Various pieces of equipment have been devised to
dilute odor samples; most dilutions are still made with syringes.

ASTM Method D1391-1957 (Standard Method for Measurement of Odor in Atmospheres--Dilution Method) gives detailed procedures.

One person can make subjective measurements of odors, especially if he has experience in correlating his reactions with those of several other people; however, a panel of observers is used for odor determinations that are likely to be presented in court or are to be used in the setting of base lines. The odor sniffers on a panel need not be trained for the work except for having a familiarity with the system to be used for identification and quantification. They should not add to their odor background with smoking, onions, perfumes, and so on, before the tests.

Chemical determinations of odors are probably best made by separating the odor fractions on a gas chromatograph, then putting the fractions from the exhaust into a mass spectrometer. As a check on this procedure and to provide tentative identification for aiding the analysis, the exhaust from the gas chromatograph should be sniffed at each peak. The GC separation and the sniffing combine to make a good method of analysis even without resorting to the mass spectrometer. The exhaust will have the pure compound smell and can be compared with the exhaust from shots of the pure compound into the gas chromatograph injection port.

Because of the difficulty in using the GC-mass spectrometer analysis and the need for analyses in laboratories that do not have a mass spectrometer, a method should be developed for making GC "fingerprints" on the emissions from various malodorous industries so that the quantification could be on the major peaks of the GC spectrum and not necessarily the peaks from the odors. The more complex analyses would be required only for a change of process and as a periodic check that the fingerprint had not changed.

I. Airborne Radioactivity

Since the advent of nuclear weapons and reactors, the determination of airborne radioactivity has become quite frequent. The radioactivity associated with particulate matter is most frequently measured, especially for samples that use concentration collection for measuring fallout. Gaseous radioactivity is used for monitoring airborne concentrations for industrial hygiene purposes in mines, reactor laboratories, and so on.

A particulate sample for radioactive analysis is collected in the usual manner for particles, most often by filtration. The filtration is often carried out on felt filters for beta and gamma analysis, but it is nearly always done with membrane filters or impaction collectors for alpha analysis. The filter, a section of the filter, or the particles collected, are put into the radiation measuring equipment for specific determination. The activity may be counted as the total radioactivity exhibited by the sample or may be broken into alpha, beta, and gamma emissions for added information. For radioisotope identifications the filter may be extracted and various chemical separations run before counting the various fractions separated. The gamma emitters may be identified by gamma spectrometry, using pulse-height analysis.

Area monitoring for radiations in the air, whether from radioisotopes or electronically produced or natural in origin, is normally done with an ion chamber for considerable amounts or with a Geiger-Müller counter for small amounts. A scintillator may be used for counting low levels of radiation, but it is more expensive and more delicate than the Geiger-Müller counter.

J. Noise Measurement

The current trend in noise measurement is toward simplification. This simplification is especially warranted in view of the number of noise measurements that will be made as a result

of the Walsh-Healy Act (regulating noise levels in industries
with federal contracts). The maximum permissible exposures
set by the act are shown in Table 7-4.

The A-weighted sound measurements have supplanted
octave band analyses in nearly all routine applications for
industrial hygiene purposes. Figure 7-2b shows the conversion
curves from octave band to equivalent A-weighted. In reading
these or similar curves, one should note that doubling the intensity
of the measured noise (pressure) adds only 6 dB to the noise level.

TABLE 7-4

Maximum Permissible Noise Exposure[a]

Duration[b] (hr/day)	Sensitivity (dBA)
8	90
6	92
4	95
3	97
2	100
1 1/2	102
1	105
1/2	110
≤ 1/4	115

[a]From "The Walsh-Healey Public Contracts Act," Noise News
Rev., 1:1, 7, February 1970.

[b]When the daily noise exposure is composed of two or more
periods of noise exposure of different levels, their combined effect should be considered rather than the individual effect of each. If the sum of the fractions $C1/T1 +
C2/T2 +...+ Cn/Tn$ exceeds unity, then the mixed exposure
should be considered to exceed the limit. Ci indicates
the total time of exposure at a specified noise level, and
Ti indicates the total time of exposure permitted at that
level. Exposure to impulsive or impact noise should not
exceed 140 dBA peak sound pressure level.

The survey meters available (a few $100) for industrial noise measurements are A-weighted in response. In acoustics and research the C-weighted (flat) response curve is often used. Formerly, recommendations were that the various weightings be used for different sound levels.

If a noise is intermittent, an integrating meter can be used which gives a "dose" value for the noise exposure. Rapid-response meters can measure impact noises of very short durations.

Octave band analyzers are rather complicated and expensive (more than $1000) to buy and to use. They are recommended only for research-type noise measurements.

K. Other Techniques of Analysis

Various other analyses have been made for air pollution in general and for specific pollutants. These efforts have involved visibility, soiling index, sulfur dioxide and humidity, and pollution specific to an area or region.

Corrosion and visibility measurements have been made as indications of pollution. Both these effects are highly dependent upon the humidity and this has led to limited success in many of the attempts at correlating the measurements with overall air pollution. Corrosion measurements are usually made by putting a piece of mild steel outside, then weighing the weight gain, or chemically cleaning the corroded material and weighing the weight loss. Corrosion has also been measured by evaporated metal films. Visibility and light scatter of atmospheric particulates have been measured by visual means, nephelometry, and laser radar (lidar).

Some efforts have been directed toward defining an air pollution index. This is hardly possible as long as the term air pollution has a vague definition. The 65 dirty cities listed

in Table 1-4 used suspended particulates, gasoline consumption, and sulfur dioxides as indicator pollutants. There are
some rather evident omissions which might be included in such
an index. These include, but are not limited to, inversion
frequency, insolation received, atmospheric hydrocarbons from
sources other than exhausts, and obnoxious odors. Another
indexing attempt used the sum of powers of sulfur dioxide concentrations and soiling index. This might be a good air pollution index for a coal-burning area, but a petrochemical complex
could register lily white on such a basis.

L. Calibration

The only assurance of accuracy for analysis is given by
calibration. Precision on replicate samples may be comforting,
but it is not indicative of accuracy. Calibrations should be made
often and the results recorded. Significant changes in calibration can indicate a serious breakdown in analytical method,
equipment, or materials.

Calibrations can be made on primary standards, or on
standards calibrated against primary standards, secondary
standards. Calibrations are normally made with the material to
be analyzed. Sometimes the calibration is made on a similar
material or against a built-in standard in the instrument.

In order for calibration to be made against the material of
test, a stock solution, standard mixture, or a quantity of known
purity must be available. Spectrophotometers often have a
simple material furnished with them for checking purposes. For
example, IR spectrometers usually have a piece of polystyrene
or similar plastic mounted as a window in a cardboard with the
correct spectrum printed on the cardboard. Chemical analyses
are usually made according to instructions which also include
directions for preparation of a calibration curve from a stock

solution. Calibrations of both the sampling and analysis require
dilute mixtures of the test material such as is found in the air.

Preparation of very dilute mixtures of various pollutant
materials has received considerable attention. The problem is
that few measuring methods are accurate over a range of 10^6,
which is necessary to obtain a 1 ppm mixture. A pollutant gas
flow of several cubic centimeters per minute requires an air
flow of several million cubic centimeters per minute. Double
dilution could be made with two serial dilutions of 1000 each.
Hersch recently reviewed methods for obtaining controlled
mixtures of pollutants. He classified the methods as mechanical
(liquid piston, common blowout, bleed from a tank, diffusion
bleeds, diffusion across a membrane, self-dilution, "pick-up
and waste"), evaporation (diffusion cell), direct evaporation,
electrolysis (generation of additive, displacement by electrode
gases, electrode gas as a vehicle of contaminant), and chemical
conversion. A common blowout uses a bottle of liquid (water)
to obtain a constant head for the air supply and the contaminant
supply which are being passed through flow restrictions which
are in the proper ratio to give the desired mixture. The method
most widely used is the diffusion across a membrane--the
permeation tube.

Apparently, any liquefiable gas can be put into a Teflon
tube with plugged ends and be made into a permeation tube. The
rate of release of the material by diffusion through the walls of
the tube is constant at a given temperature. As a result, the
amount of material released over a long period of time can be
determined by weighing the tube and its contents. Saltzmann
et al. recently devised a volumetric calibration for the amount
of material released. Tubes have been made for sulfur dioxide,
nitrogen dioxide, various hydrocarbons, halogenated compounds,
ammonia, hydrogen sulfide, hydrogen fluorides, phosgene, and
organic mercury compounds.

In addition to the calibration of methods and instruments, calibration of techniques is being attempted by having different laboratories run analyses on the same samples. This step is essential in order to determine the accuracy and the precision of the test methods.

V. SUMMARY

Air pollution analyses should be carefully planned and executed. Frequent calibrations of the analytical procedure, separately or combined with the sampling, should be made to insure the accuracy of the results. Calibrations are made against known solutions, mixtures, or amounts of the material of test.

Air pollution analyses described in this chapter include the general procedures applicable to wet chemistry, radiation, spectrophotometry, mass spectrometry, odors, and noise. Also treated in this chapter is the vocabulary necessary to present the results in the current manner. Specific procedures and methods are given for the most important tests in Chapter 9. Physical properties of particles are covered in Chapter 8.

Wet chemistry methods used are colorimetric, neutralization, precipitation, and indirect reactions. The end points and quantities may be read by color intensity, indicator, or electrochemical methods.

Spectrophotometry instruments are quite commonly used in air pollution measurements--the UV band for polycyclics, the visible band for inorganics, and the IR band for organics. Infrequently, the microwave spectrophotometer is used for free radical determinations. Atomic absorption and flame spectrometers are unique only in their energy sources. The flame spectrometer is an emission spectrometer.

SIE are undergoing a rapid rate of development at the present. For a long time such electrodes were practically limited to pH measurements. SIE are now available for many ions, but the cost of such electrodes is many times the cost of pH electrodes (approximately $200).

PROBLEMS

1. If a maximum sulfur dioxide concentration of 0.75 ppm_V was measured by an instrument with a time constant (p) of 3.5 min and the slope of the peak rise was 0.2 ppm_V/min, what was the probable actual instantaneous ambient concentration?

2. Determine the approximate geometric mean concentrations of sulfur dioxide in Fig. 1-2 and their geometric standard deviations. Plot the means with their 68.3% confidence limits against the population of the cities. What is the probable reason that San Francisco has low sulfur dioxide concentrations?

3. Make a graph for use as a calibration curve in the measurement of Fe^{3+} if the extinction coefficient (K) is 2170 liters/mole of 305 nm. Make the scales such that the slope of the line is approximately -1 as drawn (percent transmittance on log versus micrograms of Fe^{3+}).

4. Plot the curve in problem 3 as optical density versus concentration in moles per liter.

5. Why is quinhydrone used to calibrate the polarograph for oxidation-reduction potential (ORP) measurements?

6. Plot the odor intensity numbers (0-4) as ordinate versus concentration in parts per million by volume as abscissa (log) for an odor with a PIT_{50} of 1 ppm_V that becomes overpowering at 100 ppm_V.

BIBLIOGRAPHY

Air Pollution Manual, Part I: Evaluation, American Industrial Hygiene Association, Detroit, Michigan, 1960.

A. P. Altshuller, "Air Pollution," Anal. Chem., 41:5, IR-13R, April 1969.

D. M. Benforado, W. J. Rotella, and D. L. Horton, "Development of an Odor Panel for Evaluation of Odor Control Equipment," J. Air Pollution Control Assoc., 19:2, 101-105, February 1969.

J. Cholak,"Analytical Methods," Section 11, Air Pollution
Handbook, P. Magill, F. Holden, and C. Ackley, Eds.,
McGraw-Hill, New York, 1956.

E. C. Crocker and L. F. Henderson, "Analysis and Classification of Odors," Am. Perfumer, 22:6, 325-327, 356, August
1927.

L. A. Currie, "Limits for Qualitative Detection and Quantitative
Determination--Application to Radiochemistry,", Anal.
Chem., 40:3, 586-593, March 1968.

G. W. Ewing, Instrumental Methods of Chemical Analysis,
2nd ed., McGraw-Hill, New York, 1960.

B. J. Galetti and F. C. Snowden, "New Analysis Instruments
Aid Pollution Control," Environmental Sci. Tech., 3:1,
34-37, January 1969.

W. G. Hazard, "What's New in Noise?" Industrial Hyg. News
Rept., 11:1, 3-4, January 1968.

P. A. Hersch, "Controlled Addition of Experimental Pollutants
to Air," J. Air Pollution Control Assoc., 19:3, 164-172,
March 1969.

W. B. Johnson, Jr., "Lidar Applications in Air Pollution Research
and Control," J. Air Pollution Control Assoc., 19:3, 176-
180, March 1969.

A. J. Kresge and H. J. Chen, "An Error Analysis of Indicator
Measurements," Anal. Chem., 41:1, 74-78, January 1969.

R. I. Larsen, F. B. Benson, and G. A. Jutze, "Improving the
Dynamic Response of Continuous Air Pollutant Measurements with a Computer," J. Air Pollution Control Assoc.,
15:1, 19-22, January 1965.

G. Leonardos, D. Kendall, and N. Barnard, "Odor Threshold
Determinations of 53 Odorant Chemicals," J. Air Pollution
Control Assoc., 19:2, 91-95, February 1969.

J. P. Lodge, Jr. and B. R. Havlik, "Evaporated Metal Films
as Indicators of Atmospheric Pollution," Intern. J. Air
Pollution, 3:4, 249-252, April 1960.

M. J. D. Low and F. K. Clancy, "Remote Sensing and Charac-
terization of Stack Gases by Infrared Spectroscopy,"
Environmental Sci.Tech., 1:1, 73-74, January 1967.

A. P. G. Peterson and E. E. Gross, Jr., Handbook of Noise Measurement, 5th ed., General Radio Company, West Concord, Massachusetts, 1963.

Radioassay Procedures for Environmental Samples, U.S.P.H.S. Publ. No. 999-RH-27, U.S. Government Printing Office, Washington, D. C., January 1967.

G. A. Rechnitz, "New Directions for Ion-Selective Electrodes," Anal. Chem., 41:12, 109A-113A, October 1969.

J. J. Sableski, "How to Conduct a Joint Community-Wide Odor Survey," Air Eng., 10:9, 1618, September 1968.

B. E. Saltzman, "Standardization of Methods for Measurement of Air Pollutants," J. Air Pollution Control Assoc., 18:5, 326-329, May 1968.

B. E. Saltzman, C. R. Feldmann, and A. E. O'Keeffe, "Volumetric Calibration of Permeation Tubes," Environmental Sci. Tech., 3:12, 1275-1279, December 1969.

A. C. Stern, Ed., Air Pollution, Vol. II: Analysis, Monitoring, and Surveying, 2nd ed., Academic Press, New York, 1968.

P. A. St. John, W. J. McCarthy, and J. D. Winefordner, "A Statistical Method for Evaluation of Limiting Detectable Sample Concentrations," Anal. Chem., 39:12, 1495-1497, October 1967.

F. V. Wilby, "Variation in Recognition Odor Threshold of a Panel," J. Air Pollution Control Assoc., 19:2, 96-100, February 1969.

Chapter 8

PHYSICAL ANALYSIS OF PARTICLES

I. INTRODUCTION

Physical analysis of particles not only provides those particle characteristics needed to corroborate chemical identification and to design control measures, but also frequently provides positive identification without further tests; for example, cotton fibers have a distinctive appearance when viewed with a microscope.

There have been many important advances in the study of particles, including the scanning electron microscope, a rather complete catalog of particle microphotographs (by McCrone), sources of truly monodisperse aerosols, improved Tyndallometers, a filter with cylindrical holes, and an improved cascade impactor.

The physical properties that are most often desired are size, shape, cleavage, striations, optical properties, x ray diffraction pattern, density, thermal properties, and electrical properties.

II. PARTICLE SIZE DETERMINATION

A. Introduction

Particle size is most important to air pollution analysis. This can be illustrated by citing a few essential roles of particle size. Particle size is largely responsible for whether or not particles enter the lower respiratory areas of the lung and whether or not they are retained there (see Chapter 10), setting the residence time of particles in the atmosphere before they are removed, the light extinction that affects visibility and plume opacity, and the degree of removal affected by most air pollution particulate processes.

Particle sizing is deceptively difficult and requires careful attention to both the planning and the execution stages of the measurements. The proper size of sample must be collected and handled in a way that preserves the integrity of the discrete particles, with neither crushing nor flocculation.

B. Definition of Size

There are several definitions of particle diameter, which seems a logical consequence of attempting to assign a diameter to particles that are not spherical. The definitions may be divided into two types to facilitate discussion; namely, those definitions related to the size of the individual particle and those that characterize the whole population or distribution. The latter may be termed statistical diameters.

1. For Individual Particles

The most used particle diameters are Feret's, Martin's, equivalent area, and aerodynamic diameters.

a. Feret's Diameter. Feret's diameter is defined to be the length of the particle as it projects on an axis in any direction, usually the dimension across the stage (left to right) in the viewing field (see Figure 8-1). Such a dimension is the

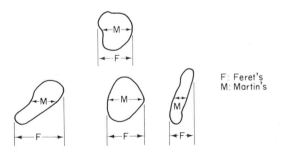

FIG. 8-1. Particle diameters. F, Feret's; M, Martin's.

one which is easiest to obtain when measuring size with an
eyepiece (filar) micrometer.

b. Martin's Diameter. Martin's diameter is the length of
the chord in a given direction (usually across the field as Feret's)
which divides the projected (viewed) area of the particle in half
(see Figure 7-1). Such a measurement is very difficult to obtain,
even on a photograph, and this diameter is not often used.

c. Equivalent Area. The area of the circle having an area
equal to the viewed area of the particle is the equivalent diameter.
The equivalent diameter is often used in particle sizing. The
reticules with calibrated circles are used for this purpose as is
the Zeiss particle counter (see p. 315).

d. Aerodynamic Diameter. The aerodynamic (or hydrody-
namic) diameter is the diameter for a sphere that would fall in
air (or water) at the same rate as the particle. This diameter
appears to be equal to the equivalent diameter because a
particle falls with a velocity proportional to its projected area
(Stokes region); however, since the two diameters are measured
by different methods and the orientation of the falling particles
in the air is not likely to be the same as that of those on the
microscope slide, two different sizes are obtained and two
different names are not inappropriate.

e. Other Diameters. Other diameters are defined in terms
of the sums or products or some other combination of the small
and large dimensions of the particles; for example,
(length X width)$^{1/2}$ or (length X width X thickness)$^{1/3}$.
Diameters are sometimes given in filar units (1 f.u. = 1.1 μm).
Unnecessary units such as this are confusing and should be
avoided. Additional diameters may be defined for other measurement methods; however, the usual practice is to calibrate
any measuring process with particles measured by microscopic
means.

2. For Populations of Particles

Statistical diameters are defined to give the best probable information, in the most compact form, about a population of particles whose size distribution is known. The term particle sizing usually refers to finding the relative numbers of particles with different sizes or within different size ranges.

Particle size distributions are usually represented by the log normal distribution (see Chapter 4). It is possible to prove theoretically (by assuming randomness of the production operation) that the size distribution of a set of particles from a crushing or grinding operation or a set from a chemical accretion or precipitation is log normal. In practice the particle size distribution obtained depends to a large extent not only upon the method of defining size but also upon the method of measuring size, especially the resolution of the method when very small particles are involved (see Fig. 8-2).

The use of a size distribution to describe a population of particles implies that a characteristic size and a degree of dispersion of sizes about that central value will be determined.

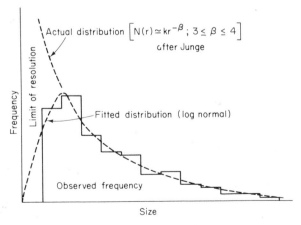

FIG. 8-2. Effect of resolution limit.

There are several different characteristic sizes variously used in air pollution analysis.

a. Common Statistical Sizes. Two sizes are used much more than others to characterize particle size distributions. These are the geometric mean diameter $(\overline{d}_g)$ and the mass median diameter(MMD).

The geometric mean size for the log normal distribution is the d_{50} size of the population, the 50-percentile size when the particles are ordered by size; that is, the $\overline{d}_g$ is equal to the count median diameter (CMD); but this is true only for this distribution. The $\overline{d}_g$ value is the diameter most frequently given, especially for measurements by any method other than dynamic. Sometimes the arithmetic mean of count size is erroneously used to characterize positively skewed distributions such as the log

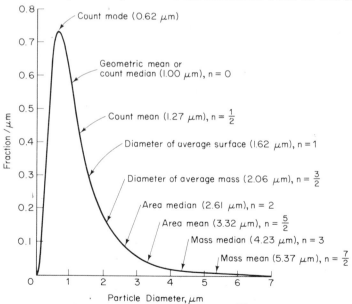

FIG. 8-3. Characteristic sizes for $\overline{d}_g = 1$ and $\sigma_g = 2$ [log normal]. Adapted from P. E. Morrow et al., Report of the Lung Dynamics Task Group for Committee II of the International Radiological Protection Commission, April 1965.

normal (Sk = 0.336; see Fig. 8-3). The relationships among
various diameters are shown in Fig. 8-3.

The mass median diameter (MMD) is that cut diameter
which would divide the total mass of the sample into equal parts.
The MMD is generally considered the best characteristic for
representing a particle size distribution when aerodynamic
properties are involved; for example, particle sizing by cascade
impactors or lung retention of particles.

b. Other Statistical Diameters. Statistical diameters other
than those defined above may be set up for specific purposes.
Normally, these additional diameters are intended to convey
more information about a specific property than either the $\overline{d}_g$
value or the MMD. One such property is the volume-to-area
ratio for the distribution, a property that is highly important in
surface chemistry governing such phenomena as dust explosions.
The average sizes are often defined by the expression

$$\overline{d}_{qp} = (\Sigma d^q / \Sigma d^p)^{1/(q - p)} ,   \hspace{2cm} (8-1)$$

where q and p give the basis of the definition; for example,
q = 1 and p = 0 for the arithmetic mean and q = 3 and p = 2 for
the average volume per area diameter. Some of the more common
specific statistical diameters are shown in Table 8-1 and need
not be further described here.

EXAMPLE 8-1: Compare the mean volume-to-surface diameter
$(\overline{d}_{vs})$ with the geometric mean diameter $(\overline{d}_g)$ for the distribution shown.

Size (μm)	f	d	d^2	fd^2	d^3	fd^3
1-1.9	16	1.5	2.25	36	3.88	54
2-4.9	40	3.5	12.25	580	42.88	1,715
5-9.9	25	7.5	56.25	1,406	421.9	10,547
10-19	14	15	225	3,150	3,375	47,250
20-49	8	35	1,225	9,800	42,875	343,000
50-99	3	75	5,625	16,875	421,875	1,265,625
	106			31,847		1,668,191

Assume log normal, $\bar{d}_g = x_{50}$.

$\bar{d}_g \approx 5 - [40 + 16 - (106/2)/40] = 5 - 9/40 = 4.8 \, \mu m$

$\bar{d}_{vs} = \sum fd^3 / \sum fd^2 = 1,668,191/31,847 = 52 \, \mu m$

c. Transformation of Diameters. Kapteyn's transformation permits the calculation of one statistical diameter from another, assuming the log normal distribution holds.

$$d_i = d_k e^{n \ln^2 \sigma_g}, \tag{8-2}$$

TABLE 8-1

Characteristic Sizes for Log Normal Distribution

Average diameter	Symbol	Algebraic definition (i = 1 to N)	n^a	Value for $\bar{d}_g = 1$, $\sigma_g = 2$
Geometric mean or count median	$\bar{d}_g$ d_{50}	antilog $(\sum_i \log d_i / N)$	0	1.00
Arithmetic mean	$\bar{d}$	$\sum_i d_i / N$	1/2	1.27
Diameter of average surface, root mean square, or surface-to-number mean	$\bar{d}_a$	$(\sum_i d_i^2 / N)^{1/2}$	1	1.62
Area median	$\bar{d}_{a50}$		2	2.61
Diameter of average mass or volume-to-number mean	$\bar{d}_v$	$(\sum_i d_i^3 / N)^{1/3}$	3/2	2.06
Area mean, Sauter mean, or volume-to-surface mean	$\bar{d}_{vs}$	$[\sum_i d_i^3 / (\sum_i d_i^2)]$	5/2	3.32
Mass median	MMD		3	4.23
Mass mean or deBroukere mean		$[\sum_i d_i^4 / (\sum_i d_i^3)]$	7/2	5.37
Harmonic mean or specific length	$\bar{d}_{hm}$	$(\sum_i d_i^{-1} / N)^{-1}$	-1/2	0.79
Specific surface or particles/area	$\bar{d}_{ss}$	$(\sum_i d_i^{-2} / N)^{-1/2}$	-1	0.62

[a] n for transformation calculation by Value = antiln $(\ln \bar{d}_g + n \ln^2 \sigma_g)$.

where d_i and d_k = statistical diameters, σ_g = geometric standard deviation, and n = characterizing exponential relation for d_i to d_k. For medians n = i - k, where i - k = 0 for count, 1 for linear, 2 for surface, and 3 for volume or mass. For means n = (q + p - 2i)/2, where q and p are defined in preceding article.

The plots of particle properties diameters d_i and d_k on log probability paper are parallel; that is, σ_g is the same for both d_i and d_k, but the sizes are displaced and $d_i \neq d_k$. For some purposes the log form of the equation is more convenient than the exponential and

$$\ln d_i = \ln d_k + n \ln^2 \sigma_g$$

or

$$\log d_i = \log d_k + 2.3026 \, n \log^2 \sigma_g.$$

Figure 8-3 shows the interrelations for the specific characteristic diameters calculated by these equations as shown in Table 8-1.

EXAMPLE 8-2: Estimate the σ_g of Example 8-1 by transformation.

$$\ln 52.4 = \ln 4.8 + 3/2 \ln^2 \sigma_g$$
$$\ln 10.9 = 3/2 \ln^2 \sigma_g$$
$$\ln \sigma_g = \sqrt{1.59} = 1.26$$
$$\sigma_g = e^{1.26} = 3.53$$

C. Measurements

Particle size can be measured by sieving, by size-dependent dynamic processes, by microscopic methods, by scattered light, by change in resistance between electrodes, and (for small particles) by the amount of charge which they will accept. The lower limits of operation for several measurement methods are shown in Table 8-2. Some sampling processes are designed to give particle separation such that the different size "cuts" can be weighed or dissolved and chemically determined. Light scattering counters normally pass a stream of air through the instrument without separating the aerosol. Most particle

TABLE 8-2

Practical Limits of Sizing Methods[a]

Particle sizing method	Lower limit (μm)
Sieving	
Standard sieves	40
Special sieves	5
Sedimentation	
Gravity with temperature control	1
without	2
Centrifuge	0.1
Goetz spectrometer	0.025
Microscopic	
Light field	0.5
Dark field	0.25
UV in air	0.1
UV in vacuum	0.005
Electron	0.01
Ultramicroscopic	0.005
Tyndallometers	0.1
Electrical charge	0.015
Condensation nuclei	0.001

[a]From various sources in bibliography.

size determinations in the past have been made by collecting the particles from a known volume of air and measuring the particle diameters with a microscope.

1. Separations

Size-selective separation of aerosol particles down to about 8 μm can be accomplished by sieving. The principles of particle dynamics are used to measure particles, or separate for measurement, by settling, centrifuging, impaction, and diffusion and migration.

a. Sieving. Collected particles may be passed through
a nest of progressively finer mesh sieves for separation. Dry
samples of 10-20 gm are placed on vibrating screens for 30 min
to $1\frac{1}{2}$ hr. The dry process is useful down to 200 mesh Tyler or
United States (74 μm), or in some cases to 325 mesh (43 or 44
μm). Water or another liquid is used to wash material through
400-mesh sieves (37 μm) and finer mesh sieves. The finer mesh
sieves are electroformed of silver and are very delicate and
somewhat expensive. They are available (Buckbee-Mears) to
5 μm. They should never be used without removal of larger-
sized particles which rapidly destroy their integrity. Membrane
filters or Nuclepore filters can be used by water-washing particles through for separation. Membrane separation is a slow
process that is often hampered by clogging of the filter, bridging
to reduce the effective pore size, and ending with a very dilute
suspension because of the large quantities of water required in
the wash. Upward filtration is preferable, but it requires pressure
feed and attention to settling rates. Such an arrangement would
put the clean water through the filter first. The same result can
be obtained by hydrometer separation, followed by filtering the
fractions from the top of the hydrometer and progressing downward.

b. Sedimentation or Elutriation. Elutriation of particles
employs their differential settling rates in a fluid to separate the
particles by size. Any fluid may be used--gaseous or liquid.
Whatever the method used it must be free of cloud settling,
hindered settling, and flocculation. Elutriation may be accomplished by fluid carry-over of particles smaller than the cut size
or by failure to settle at least as far as the cut size in a measured
time.

For fluid carry-over, a sample (often grams in size) is
agitated in a fluid stream which passes upward through a tube
with a cross-sectional area and volume flow rate combination to

give the velocity equal to the settling velocity of the cut size. The larger particles fall back into the sample holder and the smaller particles are to be carried out of the tube where they may be collected in a paper thimble or put into a larger container where they will settle out. The weight of the carry-over fractions or the loss of weight of the sample for each cut can be determined. A minimum cut size of about 5 μm is obtainable in a practical amount of time with the air elutriation equipment commercially available, and the maximum carry-over size is about 80 μm. The well-known commercial models are the Roller particle size analyzer (United States) and the Gonell elutriator (British). Electrically driven rapping hammers keep the sample agitated and off the tube walls.

Size may be measured by the sedimentation rate of particles in a fluid. The particles may be introduced at the top of the sedimentation vessel or they may be completely mixed through the fluid at the start of the test. The separation may be measured dynamically by the transmission of photons (light or x rays) through the suspension at various heights and times, by the density of the suspension with a hydrometer, by weighing the fractions left after repeated aerial sedimentations, or by weighing the settled portion after various sedimentation times. The particle size analyzer (Micromeritics) uses the attenuation of x rays to accomplish size analysis. With an electrobalance and a special accessory, cumulative sedimentation can be used to obtain the particle size distribution by recording the settled weights at various times.

Liquid elutriation is normally carried out in water, but sometimes alcohol or oil is used as the medium. A dispersing agent is usually employed to prevent flocculation. Water containing 0.1% Calgon (sodium hexametaphosphate) may be used for dusts that are insoluble in water--limestone, silica, alumina, clay, and so on; methanol is used with fresh Calsolene oil for coal, fly ash, feldspar, and so on; ethanol is used for Portland cement, gypsum, and so on. The usual procedure is to fill beakers with a solution, add a known amount of sample (0.1-1 gm or more) to the first beaker, mix thoroughly, then settle for a measured time (distance divided by settling velocity of the cut size). The liquid is decanted to the divide distance, and the supernatant transferred to a succeeding beaker. All of the particles that are transferred have a dynamic diameter less than the cut size. The liquid is replenished back to full mark and the process repeated many times for each beaker in order to remove practically all of the small-sized particles. The amount removed each time depends upon the position at which the particles started in the fluid. If the level of decantation is 3 times as far down as the distance of settling for a given particle size, the fraction removed each time should be one-half of the remaining; therefore, after six decantations less than 2% of the original fines should remain ($\frac{1}{2}^6 = 1/64$). If the volume decanted is small in relation to the overall volume in the beaker, as many as 15 to 25 repetitions may be required. The practical lower limit for liquid elutriation is 2 μm.

EXAMPLE 8-3: How long should settling be allowed to continue before decantation if 5-μm glass beads ($\rho = 3.0$) are to settle 7 cm in water?

Stokes' law: (see Chapter 5)

$$u_t = g(\rho_p - \rho_w)\, d^2/18\mu$$

$$= 980\,(3.0 - 1.0)(5 \times 10^{-4})^2/18(1.005 \times 10^{-2})$$

$$= 2.72 \times 10^{-3}\ cm/sec$$

Time = $7\ cm/2.72 \times 10^{-3}\ cm/sec = 2580\ sec \doteq 43\ min$

Very complex methods have been worked out for operating a number of beakers in series; the decanted material from one beaker is used as filler solution for the next and all of the beakers are operated simultaneously (see bibliography, Drinker and Hatch).

c. Centrifugation. Centrifugal forces up to many thousand gravities make the separation of small particles practical. The size separators fall into three categories--cyclones which bend a moving aerosol stream into a tight circle and use the velocity of the airstream as motive power, spectrometers which have slow aerosol flow through a mechanically driven, rapidly rotating head (Goetz spectrometer), and the Bahco microparticle classifier that separates a sample into size fractions by air elutriation and centrifuging.

Cyclones have been used primarily for separation of the nonrespirable fraction from the respirable dust. The cyclone collects the material that would not reach deep into the lungs. The AEC has standardized a curve for the diameter efficiency of the nonrespirable fraction as follows: 100% of 10 μm; 75% of 5 μm; 50% of 3.5 μm; 25% of 2.5 μm; and 0% of 2 μm. Cyclones are available (Unico) that closely follow these efficiencies of collection.

The Goetz aerosol spectrometer moves the aerosol through two helical channels in laminar flow while the conical head containing the channels is rotated at speeds up to 24,000 rpm. At a given speed gravities vary, because of radius variation, by a factor of about 3 and range to about 25,000. The particles are separated by size along the channels and deposited on a foil or paper or similar material for analysis. Various foils cut to shape (to form a truncated cone) are available for microscopic and chemical analysis of the deposits. Size separations can be made to about 0.025 μm, with eight ranges up to 2.7 μm. The basic instrument sells for about $3000 and ancillary equipment may double this cost.

Stöber and Flachsbart have described a flathead centrifuge with a spiral duct that can provide separation over a wide range of particle sizes in a single separation.

In the Bahco microparticle classifier, a sample of about
10 gm is moved from a hopper into spiral air currents with suitable tangential and radial velocities for separation by air elutriation and centrifugation. The airflow is varied by orifices to
change the division or cut size into coarse and fine fractions
(nine size groups). The classifier has been widely used in
metallurgical and power industries and is now receiving some
attention for air pollution work. Its principal drawback is the
large sample size required.

d. Impaction. Impaction is used to collect particles in
filtration, midget impingement, rotating rods and surfaces, and
various cascade impactors. The general features of impaction
are described in Chapter 6 (especially Cascade Impactors, p.
226) and the particle sizing aspects are briefly treated here.

The Casella cascade impactor has four impaction stages
which are usually followed by a filter. The first stage collects
virtually all unit density particles greater than 6 μm. The other
stages collect unit density particles approximately as follows:
second stage, 2-6 μm (MMD = $19\rho^{-1/2}$); third stage, 1-3 μm
(MMD = $5.8\,\rho^{-1/2}$); and the fourth stage, 0.3-1.5 μm (MMD =
$1.7\,\rho^{-1/2}$). The effect of a few large particles raises the MMD
markedly above the geometric mean.

The Unico (Lippmann) cascade impactor is calibrated from
2 to 40 liters per minute and for particle densities from 0.8 to
20 gm/cm^3 as shown in Fig. 8-4.

The Andersen sampler has six stages that collect from
1 cfm in the following approximate size ranges for unit density
particles: first stage from 8 μm up; second stage, 5-10.5 μm;
third stage, 3-6 μm; fourth stage, 2-3.5 μm; fifth stage, 1-2 μm;
and sixth stage, about 0.2-1 μm.

e. Diffusion or Force Migration Rates. Particle sizes can
be determined by separating particles according to their migra-

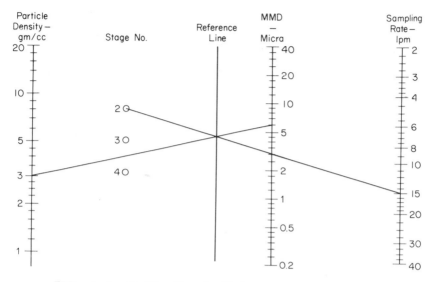

FIG. 8-4. Calibration for Unico cascade impactor.

tion rates under Brownian diffusion or under external forces, especially electrical forces. For the most accurate determinations, a thin stream containing the aerosol should be combined side-by-side with a thick stream or streams of clean air. The airstreams should have the same temperature and move with the same laminar flow velocities such that they do not mix. The purpose of a thick, clean stream(s) is to provide about the same distance of migration for all of the particles without regard to their position in the aerosol stream being analyzed. Without such an arrangement the particles collected along the path will be a mixture of sizes--large particles which started near the wall will arrive at the collecting surface with small particles that started farther away.

Diffusion is normally carried out between parallel plates or, in some cases, in a tube. Since diffusion is a random process, the Einstein equation for Brownian motion gives an expected or average distance of movement; therefore separation is not perfect and there will be size overlap. The particles collected on the wall can be removed and measured by usual chemical means or, if they are radioactive, they can be counted to determine the relative numbers of the various sizes deposited along the surface. Diffusional measurement is a difficult process and is usually employed only for very small particles (less than about 0.1 μm).

Measurement by migration of charged particles in an electrical field rests on an assumption that the charge of the particles is known. Particles can be charged by ion bombardment such that the preponderance of the particles will have a single charge, but some will have two charges and a very small portion will have more than two charges.

2. Optical Microscopy

Modern microscopy offers a wide variety of analytical possibilities. Visible microscopy is used with light field and dark field, phase contrast and interference, polarization lenses and refractive index oils, many different micrometers, graticules or reticles (disks for sizing), and microprojectors and cameras.

The basic microscope as discussed herein is assumed to consist of an eyepiece (usually 10X, although 15X is better for some applications), a barrel with focusing rack, an objective turret with 3.5, 10, 43, and 93X (or a three-position without 3.5X), a stage (which should have positioning racks for a slide), and a substage (Abbe) condenser with iris diaphragm.

a. Resolution. Particles too small for viewing with the unaided eye are most often sized, counted, and so on, with optical microscopy. The lower limit of resolution (ability to separate small details) depends upon the numerical aperture of the objective. The numerical aperture is defined as

$$N.A. = n \sin \frac{\alpha}{2},$$

(8-3)

where N.A. = numerical aperture of the objective, α = angular
aperture (see Figure 8-5), and n = lowest index of refraction
between focal plane and objective lens. The numerical aperture
of the best oil immersion objective lens is about 1.3. The
theoretical limit of resolution is related to the wavelength used
to view the object, usually about 450 nm for optical microscopy.

$$s = \frac{1.22\lambda}{2 \, N.A.} , \qquad (8-4)$$

where s = distance between the points at limit of resolution and
λ = wavelength of light used to view the points.

EXAMPLE 8-4. What is the normal lower limit of resolution for
the conditions which have been described above for an
oil immersion lens and optical microscopy?

$$s = \frac{1.22\lambda}{2 \, N.A.} = \frac{1.22 \, (0.45 \, \mu m)}{2(1.3)} = 0.21 \, \mu m$$

b. Light Field vs Dark Field. In light field microscopy
light rays from the substage lamp or other light source are directed
by the condenser through the slide and into the objective lens.
The silhouettes of the particles that stop or bend the light rays
are viewed against the light background. The dark-field condenser passes light rays from the periphery of the lens through
the field in a direction such that the rays are not picked up by

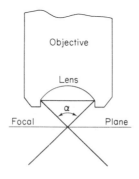

FIG. 8-5. Angular aperture (α).

the objective. The particles scatter the light rays into the
objective and are seen as bright light sources against a dark
background.

Counts made with the dark field show more small particles
than those with the light field; however, small particle measurements with the dark field may actually be more difficult and less
accurate than those with the light field. Dark field photomicrographs are especially striking for refractive particles such as
water droplets or quartz particles.

c. Microprojection and Photomicrography. The tedium of
counting and/or sizing particles with a microscope can be re-
lieved to a large degree by the use of microprojection or photomicrography.

A microprojector is often used in counting and sizing
particles to project a microscopic image through a prism over
the eyepiece onto a frosted glass or a white enameled screen.
The projected image is several inches in diameter. Microprojection is done in a darkened room and uses a rather intense
light source.

Photomicrography provides a phtographic record of the
field of view. This has the advantage that a particle counter,
such as the Zeiss, can be used to determine the statistical
diameters of choice. Furthermore, a pin prick can identify those
particles that have been counted and will help avoid duplication
errors. It has a disadvantage in resolution of the smallest
particles, but modern films have lessened this drawback greatly.
These films also have speed ratings (ASA numbers often greater
than 1000) that permit photomicrography without long exposure
times.

d. Microscopic Accessories. There are numerous particle
size-count accessories for the optical microscope. Some are
practically essential and others are labor-saving conveniences.

A stage micrometer is almost a necessity. The usual ones are the same size as a slide (1 x 3 in.) and have a polished section in the center. At the center of the polished section is a 2-mm (2000-μm) scale which is marked into divisions, the smallest division being 10 μm. Such stage micrometers are available from laboratory supply houses for about $25.

Various types of graticules or reticles have been devised to ease the task of size counting. These graticules are marked on disks which insert into the eyepiece between the lenses. The most popular graticule is probably the Whipple disk (about $20), but the Cawood-Patterson globe-and-circle is often used, and sometimes the Fairs or other forms are the choice (see Fig. 8-6).

There are two types of eyepiece micrometers in common usage in size counting, the filar micrometer and the image splitter (Vickers). An eyepiece micrometer replaces the regular eyepiece. In the filar micrometer the hairline is moved from one side of the particle image to the other with a calibrated thumbscrew. The image splitter also uses a calibrated thumbscrew but moves separate images of the particle from tangency to identical overlap or vice versa. There are built-in filters to show one image in red and the other in green; then the superimposed image is in yellow. An electric flip-flop device is available to give a go-no go type of measure for counting the particles within set boundary sizes.

The success of optical microscopy depends upon the proper lighting--lighting that gives good contrast between the particle and its background or will give adequate strength for polarization, interferometry, microprojection, or microphotography. The lighting may be as simple as a substage lamp or even a mirror for reflection of daylight or room light into the substage condenser, or it may be as complex as the orthoilluminator (about $600) for phase contrast work. The usual substage lamp has a 15 W bulb

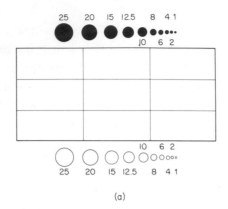

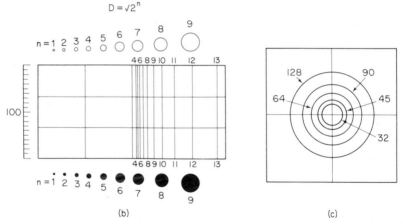

FIG. 8-6. Graticules. (a) Patterson-Cawood. (b) Porton-Fairs. (c) Fairs No. 3. From K. R. May, "The Cascade Impactor: An Instrument for Sampling Coarse Aerosols," _J. Sci. Inst._, _22_:10, 187-195, October 1945.

with frosted blue glass and fits directly under the substage condenser. The Köhler illuminator focuses the filament of a clear glass bulb in the plane of the iris diaphragm; therefore the lamp must have a condenser and a diaphragm. This illuminator is capable of a higher intensity than the substage lamp (about 75 W). For microprojection or some camera work, an even stronger light source may be desirable. One such is the carbon arc illuminator. Particles must not be too temperature sensitive or long-term illumination by strong sources changes their shape (especially biological particles) or size (vaporization). The illuminator may have filters built into it to obtain any color of light or polarized light or, in some cases, the filters are built into the microscope.

The dark field accessory includes a dark-field condenser and a built-in variable intensity light.

Crossed Nicols are needed to measure some of the optical properties and to obtain colors in birefringent particles. The polarizing filter which gives unidirectional light to the sample is followed in the barrel of the scope by a polarizer at 90° to the first. No light comes through unless it is bent by the particles viewed. The polarizer-analyzer pair can be rotated to give colors.

The most popular counting cells are the Sedgwick-Rafter, the Dunn, and the hemocytometer cells. The Sedgwick-Rafter cell is a slide with a rectangular, 1-ml reservoir (1 mm deep X 10 cm^2 area) which is formed by a raised flat rim. The rim is greased, the reservoir filled, and a special cover glass set over the reservoir. The Dunn cell has three pieces--a base plate (slide), a rectangular section 1 mm thick with a round hole for the sample, and a cover plate. The parts are sealed together with grease and are disassembled for cleaning. The hemocytometer uses a 0.1-mm-thick sample in a cell that has a ruled bottom. This cell, which was designed for blood counts, eliminates the need for an eyepiece graticule, but it is rather expensive for student use since it is often broken.

The Zeiss particle counter is an instrument for size counting of particles, especially by equivalent area diameter. A leaf diaphragm is adjusted to provide a light spot of area equivalent to the photographic image ; then a pedal is tripped to register

the diameter of the diaphragm opening. A bank of registers
record the tally by size groups (up to 48 plus a total), which
may be set to increase linearly or logarithmically. The device
punches a pin into the particle image to prevent recounting. It
sells for about $3500.

Some sampling microscopes have an impactor sampler
built into a microscope. These instruments emphasize portability.
The impact disk or collector may be moved to new positions for
successive samplings of the same or different dust population.
Two samplers of this type are the Zeiss konimeter and the Owens
jet dust sampler.

Various other accessories or supplies for microscopes make
size counting possible or make it easier. The major items have
been described. Some of the minor items are slides and slide-
holders, cover glasses, tally counters, a photoilluminator, a
transparent overlay grid, and dust skirts or hoods to keep the
background count low. Two sizes of slides are commonly used,
the standard 3 in. X 1 in. X 1 mm and the 3 in. X 2 in. X 1 mm.
The latter is needed for mounting 47-mm or 2 in. membrane
filters. Both round and square cover glasses are convenient
for different purposes. The former are used in the Casella cascade impactor for the collection of particles. Tally counters
are a real convenience for particle counting because they permit
the tallying of particles in selected size groups without looking
away from the field. One with 10 group counters, 2 rows of 5
each, works best for particle size counts. In the event that more
than 10 groups are desired, the few particles in the largest groups
can be counted before the tally starts. A clear plastic overlay
with grids or other patterns is needed for photograph counting if
the Zeiss particle counter is not being used. When large numbers
of particles are shown, only part of the sections need be counted.
A washable ink pen can be used to mark particles as counted
without permanently defacing the photomicrograph.

e. Sampling for Particle Size Counting. Whatever sampling method is used, cognizance should be given to obtaining about 75 to 300 particles in the microscopic field used. For liquid collection, dilution is readily carried out, and in some extremes evaporation may be. The dry methods of collection give fixed particle density, but the particles can sometimes be put into a liquid suspension. The sampling time can be varied to obtain the proper number of particles on a dry sampler. The sampling should be done isokinetically (see p. 238) except for small (long-residence) particles.

f. Sample Preparation for Particle Size Counting. The sample collected directly on slides by sedimentation, thermal precipitation, or cascade impaction may require no preparation. If it is desired to preserve the samples after viewing (particularly for oil immersion viewing), the particles should be fixed with collodion (Plexiglas dissolved in ethylene dichloride), or a similar substance. Alternatively, a cover glass may be fixed over particles and sealed with Vaseline or stopcock grease.

Particles collected in liquid or washed into liquid are counted in cells or by evaporation of liquid on a slide. If size distribution and not particulate density, millions of particles per cubic foot (mppcf), is needed, a drop of liquid suspension may be put on a slide and flattened by a cover glass. Some differential particle migration may occur and fields near the edge and near the middle of the spot should be counted. When counting cells are used, a portion of the liquid is put into a clean cell, covered with a cover glass (sealed with Vaseline), and allowed to settle, 30 min for water or 20 min for alcohol, for 1 mm depth. The particles must be in one plane (within the depth of field) for focusing.

Membrane filters are rendered transparent by many oils and solvents. If it is not desired to preserve the filter sample, the

filter may be placed on a slide, face down, and immersion oil
or other oil used to make the filter transparent. If it is desired
to save the sample after counting, place the filter on a slide and
wet with cyclohexanone, dimethylformamide-ethyl alcohol mixture, diethyloxalate-dimethylphthalate (1:1) solution, or any of
several solvents. It is quite important to prevent washing of the
particles which segregates them by size. The best method is
to apply the solvent by condensing from the vapor or, secondarily,
to apply the solvent in small droplets with an atomizer. If
segregation is apparent, the mounted material should be dissolved
with acetone or Cellosolve and the solution reapplied to the slide
after thorough mixing. Filters with too many particles for counting
may be dissolved in acetone and diluted as much as desired.

A section of sticky tape pressed against a dusty surface
for sampling may be dissolved in a solvent to give liquid suspension for mounting, or it may be fastened directly to a slide.

g. Particle Size Counting. The microscope should be in
a dust-free room. If the room is not reasonably dust-free,
fashion a skirt from plastic (sandwich bags or something similar)
and attach so that it covers the lower portion of the microscope.
The eyepiece should contain a graticule to facilitate counting--
a Whipple disk for linear measure or a graticule with circles for
equivalent diameters. A microprojector and a good light are
desirable.

Select the proper magnification for the job. Use the lowest
magnification that gives the resolution needed (see Table 8-3).
For example, if it is desired to group the sizes into particles
> 5 μm, a magnification of 100X (10X by 10X) could be used for
the first group and 430X (10X by 43X) for the second group, but
a correction for the relative areas viewed would be necessary.
As a result, the 430X would normally be used for all such counting. If it is desired to size-count all particles possible using

TABLE 8-3

Magnifications Required by Various Sizes

Magnification	Lower size limit (μm)	
	For counting	For detection
30X	30	A few micrometers
100X	5	1
430X	2	0.7
970X	0.5	0.3

optical microscopy, the last group will be counted with the oil immersion objective (970X, or 10X by 97X).

The particles should be tally-counted by size, normally in 8 or 10 groups, and the data prepared on a graph to determine the distribution; likely a log probability plot will be used. A background count of a filter or cell that has received the same treatment, except for the sampling, should be subtracted from the sample count. A minimum of 4 background counts per 100 samples is desirable. The background frequency plot should be smoothed and the number from the smoothed plot subtracted. This procedure puts the statistical errors from a small sample at a probable minimum.

A technique called multistage sampling has been recommended by Whitby, and adopted as a standard by the Pennsylvania Department of Health. The first count takes in all of the size groups (that is, 8 groups) and covers more than 200 particles. The second count omits the counting of those particles in the first group but assumes that they are present in the same proportion as before; that is, the number for group 1 in the second count is calculated by multiplying the number present in group 1 in the first count by the ratio of the total count in groups 2 through 8 for the second count to the same groups for the first count. The process is repeated, and groups 1 and 2 are calculated in the

next counting operation from the ratio of the number in groups
3 through 8 in the third count to the number for the first two
counts. The operation is continued until all the groups have a
sufficient number of particles for statistical accuracies. The
ratio calculation should not be carried far enough to involve a
small number of particles because it would then lack statistical
significance.

The membrane filter normally used for particle size work
has an imprinted grid to facilitate counting. The squares are
3.08 mm to a side, such that there are 100 squares in the
effective filtration area of a 47-mm filter. The filter should be
oriented on the stage for the stage movement of the slide to
follow along a grid line. Considerable movement of the lines
upon dissolution of the filter for viewing is a good indication
that the particles have been washed about also.

A random field should be selected, and the number of particles in that area estimated. A decision as to counting all the
particles in the graticule field must be made. A total count of
more than 200 is necessary, and one of several hundred is
desirable. If the graticule ruled space is estimated to contain
300 particles, perhaps a decision to count one-fourth of the
field in five different fields will be made. The selection as to
which sectors of the grid are to be counted should be made and
the five fields selected at random without looking at the field
during the decision. At least 10 to 20 particles should be
counted in any size group that significantly affects the distribution as related to the use to be made of the data.

EXAMPLE 8-5: How many particles are contained in 1 ml of a
liquid suspension if the microscopic count of five Whipple
disk fields on a Sedgwick rafter shows 480 and each of
the large squares of the disk is 100 μm on an edge?

Size of field: 10 squares X 100 μm/square = 1000 μm

Number of fields: Area of cell = 10 cm^2

$$\text{Area of field} = 10^6 \ \mu m^2 \approxeq 0.01 \ cm^2$$

Number of particles per millimeter: $\dfrac{10 \ cm^2}{0.01 \ cm^2} = 1000$ fields

$$1000 \text{ fields} \times 480/5 \text{ fields} = 96{,}000/ml$$

3. Electron Microscopy

In electron microscopy an electron beam is magnetically focused on a sample mounted on a material that is transparent to electrons; the divergent beam is absorbed on a fluorescent screen for viewing or photographing the silhouette of the object in focus. The beam must be in a hard vacuum to prevent air scatter of the electrons. The electron microscope can resolve particles down to about $0.01 \ \mu m$ by normal means. New techniques and equipment continue to lower this size limit.

Samples are often collected by electrostatic precipitation on a collection film which is either supported on, or later mounted on, a small (about 1/8-in.) circle of 200-mesh copper wire. There are many other techniques for sample preparation; for example, a carbon film deposited on a glass slide may be used for thermal precipitation of the sample, then the film floated off the slide and onto the wire mesh. The carbon film can be put on the slide by carbon arc within a vacuum jar.

The scanning electron microscope makes numerous passes and scans across a particulate sample. By this means an extremely large depth of field is obtainable. The scanning electron microscope gives by far the most detailed information that has ever been available for small-particle configurations. The micrograph obtained is practically three-dimensional (see Fig. 8-7).

4. Light-Scattering Methods

The amount of light scattered by a particle is often used as a measure of the size of the particle. The particle sizes of most general interest in air pollution (about $0.1-10 \ \mu m$ or more) are best measured by forward scattered light--the Tyndall effect.

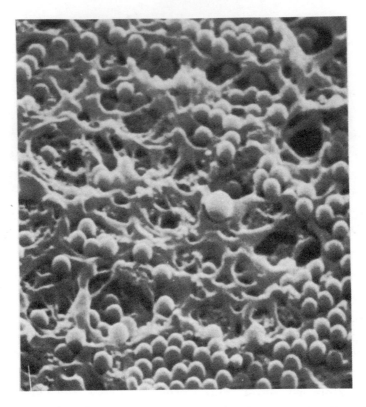

FIG. 8-7. Scanning electron photomicrograph. From E. Bulba, "Scanning Electron Microscopy," Dust Topics, 6:1, 4-9, February 1969.

The operation of Tyndallometer instruments may be described briefly (see Fig. 8-8) as follows. A light source is focused through an aerosol stream; a photomultiplier tube pointing through the focal cone at a blacked-out place on the light-focusing lens picks up light from particles; and a pulse-height analyzer measures the size of the photomultiplier pulse. The pulse-height analyzer has been single channel (counting one size each period) but now multichannel instruments are being used more (two to hundreds of sizes being counted at one time).

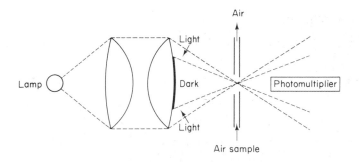

FIG. 8-8. Principle of Tyndallometers.

Tyndallometer-type instruments are manufactured by several companies, with prices starting at about $3500. They are very popular and are now found in nearly all clean room installations, although there are some drawbacks in the use of such instruments. Particles in the sensitive volume at the same time are counted as one large particle. The amount of scattered light depends somewhat upon the type of particle as well as its size. Relative humidity, especially above 70%, changes the measured size. However, there are errors in any measuring system and the ease of application makes these counters attractive.

The oldest use of light scattering for particle counting and/or sizing is nephelometry. The nephelometer merely measures the amount of light attenuation from the particles in the light beam. If all of the particles are of a given size, or if an effective size is known, the attenuation can be interpreted as relating to the number of particles present. Condensation nuclei counters detect small particles (0.001-0.1 μm) by coating them with water to make them large enough for light-scattering measurements. The corona method may be used to measure particle size in the range 10-30 μm. The diffraction rings are positioned according to Fraunhofer diffraction.

5. Electrical

Electrical counters are of three types--electrical resistance change, amount of charge accepted, and piezoelectric. The Coulter counter was one of the earliest particle counter-sizer instruments. In this type of instrument, the particles are put into a liquid electrolyte suspension and passed between electrodes. The amount of change in the current is a function of particle size. Current changes are analyzed in a window-type, single-channel analyzer. This instrument is very good for counting bacterial cells, blood cells, and other particles normally in a liquid suspension and having a usual size that does not vary widely. It is also quite usable on particle distributions from the atmosphere if they can be put into liquid suspension.

The amount of charge that a given particle accepts is a function of its size. Millikan used this principle in his oil drop experiment to measure the electronic unit of charge. Such an instrument is applicable in the particle size range of 0.015-1 μm (see Whitby).

The photoelectric detector has been developed for counting and sizing micrometeroids. The particle impacts on a photo-electric crystal at very high velocity. This method is being used in space applications, but it needs further development to increase its sensitivity to small particles for use in air pollution work. Practical velocities limit the minimum detectable size to the order of 10 μm.

6. Other Measuring Methods

A layered filter can be used to determine particle size by the number of layers (depth) penetrated by the particle before collection. The determination is made easier by use with radioactive particles.

Acoustical counters have been devised to give a pulse when a particle passes through the sensor capillary in front of a micro-

phone. More developmental work will be needed if this device is
to be made practical.

7. Surface Area

The surface area of particles can be determined by calculation from diameters measured and a shape factor or from mass
and diameter measurements, by the amount of gas required to
cover the particles with a monolayer of molecules, by the permeability of a sample to flow of a fluid, by the light extinction
caused by the particles, and by the heat of wetting or heat of
solution.

The results of measurements by the different methods vary
markedly. Herdan cites four samples of a molybdenum powder
analyzed by the most popular methods with results as follows
(in square centimeters per gram): microscopic diameter, 440
(220-470); permeability, 1500 (1040-1810); and adsorption, 3400
(3110-4010). The differences are attributed to the adsorptive
measurement of total area of fissures and outer surface, the flow
permeability measurement of the outer surface only, and the
microscopic measurement of agglomerates as single particles.
The method selected should be in line with the use of the data,
or a correction should be made to change the results to another
compatible system.

8. Particle Volume and Pore Size

By using the proper type of fluid and manipulating the
pressure and the pressure release, it is possible to measure the
volume of particles and of pores in particles. Orr has developed
the use of mercury penetration for determination of volume, pore
size, and even pore shape.

D. Errors in Size Counting

The particle size count desired is obviously the one that
affects the process of interest. If settling properties are the

parameters of interest, care should be exercised not to crush or
tear apart flocculated or agglomerated particles. If, however,
adsorptive properties are under investigation, the discrete particles rather than their agglomerates should be measured. In
the first case thermal precipitation, electrostatic precipitation,
or a sedimentation chamber should be used for collection, while
in the second case an impaction method of collection may be
desirable. This illustrates the importance of knowing the purpose of data collection before selecting the sampling method.
Errors in size counting should be defined as failure to obtain
the size count that accurately determines the aerosol properties
of interest. Crushing, flocculation, and agglomeration all contribute to errors at some time. The extent of these errors is yet
to be determined for most methods.

The elision of the lower end of the distribution curve by
method limitation should not necessarily be considered an error.
If the missing portion of the curve does not affect the process,
either because the mass of the missing part is insignificant or
judgment is based on past experience with the elided distribution
curve, the omission of that portion of the distribution is purely
academic; however, cognizance of the nature of the true distribution curve often prevents mistakes in application.

Particles collected and prepared for counting may overlap
in the field of view. The higher the particle loading is the more
probable it is that overlap will occur. A correction recommended
by Davies et al. is the following:

$$N/N_o = -A/(N_o \pi \overline{d^2}) \ln (1 - N_o \pi \overline{d^2}/A), \qquad (8-5)$$

where N = true count, N_o = observed count, A = area of deposition, and $\overline{d^2}$ = mean square diameter of all particles.

EXAMPLE 8-6: What is the correction factor (N/N_o) for 20%
coverage of viewing area?

$$N/N_o = - 1/0.2 \ln (1 - 0.2) = -5 \ln 0.8$$
$$= -5 (-0.223) = 1.12$$

Some analysts have claimed that large errors in size count measurements are unavoidable. Such standard errors as 50% (of the larger count) between observers and 33% by one observer have been quoted as to be expected. Reasonable care by a conscientious, careful person can keep the measurement error by a single observer to not more than ±10%, and utmost care can keep it to about ±5%. The sampling error should not exceed 10% for good sampling techniques. The combined sampling and measurement errors would then be about 14% $[\sqrt{(10^2 + 10^2)}$ see Chapter 4, Section XII]. Individuals can be checked by having them run a size count on a standard sample or by inclusion of more than one portion of a single sampling collection for a duplication check.

Systematic errors may be made through erroneous calibration of a measuring device or system. Such errors will be discovered if independent checks by two or more observers are made on a single sample.

E. Calibration

The calibration of particle size count methods is becoming much easier since the advent of monodisperse particles. There are also more sophisticated electronic checks of some measuring system.

1. Monodisperse Aerosols

Monodisperse particles are all the same size ($\sigma_g = 1.00$, or $\sigma = 0.00$). Plastic particles from the Dow Chemical Company are truly monodisperse, within the bounds of today's definition of the term. Particles 0.09-1.10 μm are polystyrene latex; those 1-6 μm are polyvinyl toluene latex; and those 6-100 μm are styrene divinylbenzene copolymer. The 95% confidence limits on the diameters are usually less than 0.012 μm. The particles are received in a liquid suspension. They are used by nebulizing the suspension into a chamber where evaporation leaves the particles in the air. Doublets and triplets may be removed by impaction through an orifice onto a surface. Doublet and triplet production may be lowered by dilution of the suspension, but there is a limit to the use of this technique.

Another source of monodisperse particles is an exploding wire. The particles produced by an exploding wire are in a narrower size range than that for the Dow particles (generally from a few hundredths to a few tenths of a micrometer). A conducting wire is exploded by discharging a large-capacity condenser through the wire. This produces a loud noise and a cloud of smoke. The wires that have been used include aluminum, silver, gold, platinum, iron, copper, and others.

Prior to the advent of the Dow particles, the only nearly monodisperse particles available were spores of certain plants and microorganisms (for small sizes) and pollen, glass beads, or metal shot (for large sizes). These were not truly monodisperse, and several methods were devised for obtaining such particles. DOP smoke generators can give useful particles and they have the advantage of having been widely used. Methylene blue and uranine aerosols have been generated by various means, but primarily by smoke generators, spinning disks, and nebulizers (aspirators). DOP smoke and methylene blue often have a geometric mean size of about 0.3 μm and uranine may have an MMD as small as 0.025 μm.

2. Polydisperse Particles

Calibrations have been made using admittedly polydisperse particles such as powders from the chemical shelves of the laboratory. Some particles used for this purpose are cupric oxide and calcium carbonate. The particles give a good range of sizes for testing certain air cleaners and for optical counting.

3. Other Calibrations

Electronic counters, such as pulse-height, light-scattering counters, may have a built-in source or source equivalent resistance to set the signal strength for calibration. Different particles have different albedo values, and the equivalent resistance may need to represent particles of various sizes for varying particle types. The forward scattered light (Tyndall effect) is the most reliable path of the light-scattering measurements.

III. OTHER PETROGRAPHIC PROPERTIES

In addition to size, area, volume, and pore volume, there are several other physical properties that can be used in particle analysis. These include appearance, optical properties, density, melting point and freezing point, vapor pressure and boiling point, dielectric constant and x-ray diffraction properties.

A. Appearance

The appearance of many particle types when viewed with the microscope is distinctive enough to establish positive identification. The properties to look for are size, shape, cleavage, striations, albedo, color, and transparency. Cotton fibers, soot particles, quartz dust, pollens (various), viscose fibers, silk fibers, mica, and many other types of particles can be identified with relative certainty by comparison with slides of these substances or with photographs of their microscopic appearances (see Fig. 8-9). There are sets of slides available for the common particles, and there is an extensive catalog of photomicrographs for comparison purposes (The Particle Atlas). The color and appearance may be more distinctive under polarized light or with interference microscopy. The angle of lighting can be used to measure the thickness of particle and surface irregularities from the shadows cast by the particle.

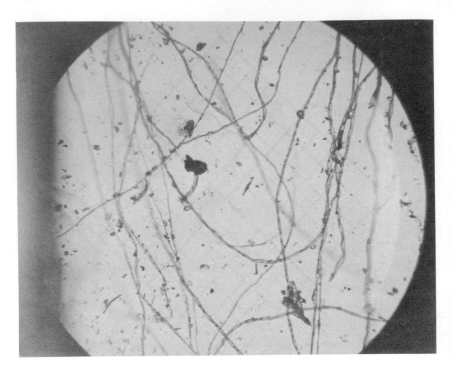

FIG. 8-9. Cotton fibers and room dust.

B. Refractive Index

The refractive index and its related optical properties can aid significantly in the identification of particles. The spacing of atoms within crystalline materials determines their index(es) of refraction. The refractive index (m) of a material is defined as the ratio of the velocity of light in a vacuum to the velocity of light in the material. The approximate refractive index of air is 1.00029, of water 1.33, and of glass 1.50. The refractive index depends upon the wavelength of the light (usually sodium light, $\lambda = 5893$ A, is used) and the temperature of the medium (normally given at $20^{\circ}C$ and changes about $0.0005/^{\circ}C$).

Transparent particles disappear under a microscope if the m value of the particles equals the m value of the liquid surrounding the particles. Calibrated liquids, usually oils, are available for measuring m values. The m value of a liquid can be changed by mixing oils of different m values, or small changes can be made by temperature adjustment. McCrone notes that the use of temperatures -75 to 125°C with three liquids (Canada balsam, Aroclors, and methylene iodide) gives m values from 1.48 to 1.82, a range containing more than 95% of atmospheric particles.

In practice, the m values of particles are often found by bracketing with two oils and making use of the Becke line phenomenon. The Becke line is a bright halo near the edge of a transparent particle. It shifts toward the higher m value (particle or liquid) as the microscope is raised above the focal point; therefore the particles with m values greater than that of the liquid appear bright. For example, quartz is identified by using oils with m values just below and just above 1.544. The difference between the number of bright particles in the two oils when the lens is just above the focal point indicates the number of quartz particles.

Materials with a single value of m are termed isotropic, and their crystalline structures cubic. Some plastics, glass, and several compounds exhibit this type of structure. Anisotropic materials show two or three principal indexes.

C. Other Optical Parameters

There are several optical properties other than index of refraction that are often used in particle identification. Some of these are related to the index of refraction. The most frequently used property of this type is birefringence. Birefringence (difference between the largest and smallest m values) causes colors in particles viewed with crossed polars. The light is polarized before reaching the particles; then an analyzer is placed with 90° orientation to give a black field of view. The color seen depends upon birefringence and particle thickness. Other optical properties are related to the index of refraction, such as the elongation and the extinction angle.

D. Density

One of two general approaches is usually employed to measure density of particles according to whether a ponderable sample or a number of particles is used.

The density measurement on a large sample of particles is made by enclosing the weighed sample in a cell of known volume, measuring the volume of sample by the amount of fluid space occupied by the particles, and calculating the density. Various gases, mercury, and other materials are employed to measure the volume of the particles. The pores may or may not be included as part of the particle volume.

The density of individual particles is often measured by their settling rate in a fluid. A particle can be put into a tube filled with layers of varying density liquids and its resting place observed, or it can be put into a tube with a liquid which is heated until the particle starts to settle. The density of the liquid with temperature must be accurately known. The density determination can be made from settling rate in air, but a liquid is used more often. The separation of particles by density can be carried out by increasing the density of the liquid (adding and mixing a heavy liquid) and decanting the top portion with the floating particles. For example, carbon tetrachloride (density, 1.6) can be mixed with methyl iodide (density, 3.15). The density column can be calibrated by using known density particles.

Consecutive photomicrographs of settling particles can be used to determine the settling velocity. With the size, velocity, and viscosity of the medium all known, the density can be calculated.

E. Thermal and Other Related Properties

Thermal and other related properties can be quite useful in identifying particles that are pure compounds or elements, and knowledge of such properties is often required for the design of air-cleaning systems. The most important properties for individual particles are melting point or freezing point, boiling point-vapor pressure, and ignition temperature. The thermal conductivity and specific heat of the particles in bulk with the entrained air are usually more important to design than they are for the individual particles.

The freezing point or melting point may be measured by use of a cold- or hot-stage microscope accessory. Particles broken with a dissecting needle will be round if liquid and angular if not.

Vapor pressure can be measured by floating particles on a clean column of mercury in an inverted buret. The mercury and the particles should be free of gas, and the measurement must be corrected for the vapor pressure of the mercury. There are more accurate methods for vapor pressure using more refined equipment and technique.

The boiling point can be measured by the temperature required to give a vapor pressure equal to the atmospheric pressure, or by heating the melted particle mass until the temperature levels off.

A differential thermal analyzer can be used to measure thermal conductivity, specific heat, and heats of reaction. The equipment is simple in concept (an insulated, heatable container for the sample with thermocouples to sense the temperatures in the sample), but its uses and capabilities are really sophisticated.

F. Other Physical Properties

The electrical and magnetic properties of particles may aid in their identification, but the principal significance of these properties is in particle collection design. The electrical properties that can be measured include the amount of charge the particles (of a given size) will take (the dielectric constant) and the resistivity or conductivity of the particles (singly or in bulk). The magnetic susceptibility of the particles may also give a clue to their identification.

x-Ray diffraction, emission, and fluorescence may be used in a manner analogous to the optical properties of the particles. Single particles may be mounted for study, or the bulk of a deposit may be used.

IV. SUMMARY

The physical properties of particles are being used to identify particles and to aid the design of collecting apparatus. Positive identification of many particles can be made from microscopic examination alone. Computer-aided machine identification by scanning electron microscopy is possible with many particles such as pollens, microorganisms, and certain crystal shapes. Automatic particle size counting is beginning to emerge. Most air-cleaning methods have size-dependent efficiencies, making a careful size distribution determination mandatory.

Size counting of particles is done by several means, but the usual method, and the referee method, is microscope viewing. The light microscope permits viewing of the particle silhouette of particles larger than a few tenths of a micrometer in diameter against a light background (light-field microscopy), or the particle image from the scattered light against a black background (dark-field microscopy). There are many convenient accessories to aid in size counting, including stage micrometers, eyepiece micrometers, graticules, microprojectors, Sedgwick-Rafter

counting cells, blood count cells, and others. Only the stage
micrometer and the graticule could logically be called essential.
The electron microscope produces an image by electrons rather
than light rays. It can resolve particles to 0.01 μm or less.
The scanning electron microscope provides a most satisfying
view of particles.

Area, volume, and pore size are measured by the use of
fluids with pressure and/or vacuum. Adsorption of an assumed
monolayer of nitrogen or other inert gas, or the permeability to
gas flow, may be taken as a measure of surface area. Volume
and pore size are measurable by using mercury and pressure
variations.

Other physical properties that aid in the identification or
collection of particles are appearance, optical properties, x-ray
diffraction patterns, density, melting point, freezing point,
boiling point, vapor pressure, and electrical and magnetic properties.

PROBLEMS

1. What is the probable orientation of the minimum dimension
   when viewing nonspherical particles with a microscope?
2. What is the arithmetic mean size for the distribution of
   Example 8-1 data?
3. What is the value of the MMD $(\overline{d}_m)$ for the CMD equal to
   1.0 μm and σ_g equal to 3?
4. What are the geometric mean diameters $(\overline{d}_g)$ for the MMD
   values shown for the Casella cascade impactor if σ_g
   equals 2?
5. Devise a method to allow simultaneous decantation of water
   for the minus 5-μm and the minus 2-μm beakers. Is any
   other method possible?
6. What velcoity obtains in the drum of the Roller analyzer to
   carry over particles less than 10 μm in diameter with ρ
   equal 2.65?
7. What would be the maximum radius of rotation for the Goetz
   spectrometer to obtain 25,000 g's at 24,000 rpm (p. 307)?

8. Why should resolution with UV light be lower than for visible light?

9. What are the 90% confidence limits for the 96,000 particles per cm^3 of Example 8-5 if σ equals approximately the square root of the number counted?

10. What is the mean square diameter $(\overline{d^2})$ of Example 8-1 data?

11. What is the observed count (N_0) that agrees with Example 8-6?

BIBLIOGRAPHY

Air Sampling Instruments, 3rd ed., American Conference of Governmental Industrial Hygienists, Cincinnati, Ohio, 1966.

B. Binek, B. Dohnalova, S. Przyborowski, and W. Ullmann, "Using the Scintillation Spectrometer for Aerosols in Research and Industry," Staub, 27:9, 1-7, September 1967.

E. Bulba, "Scanning Electron Microscopy," Dust Topics, 6:1, 4-9, February 1969.

R. D. Cadle, Particle Size, Van Nostrand Reinhold, New York, 1965.

M. Corn, "Statistical Reliability of Particle Size Distributions Determined by Microscopic Techniques," Am. Industrial Hyg. Assoc. J., 26:1, 8-16, January-February 1965.

P. Drinker and T. Hatch, Industrial Dust, 2nd ed., McGraw-Hill, New York, 1954.

G. H. Fricke, D. Rosenthal, and G. Welford, "Density Distribution of Solid Particles by a Flotation-Refractive Index Method," Anal. Chem., 41:13, 1866-1869, November 1969.

G. Herdan, Small Particle Statistics, 2nd ed., Butterworths, London, 1960.

R. R. Irani and C. F. Callis, Particle Size--Measurement, Interpretation, and Application, Wiley, New York, 1963.

J. O. Irwin, P. Armitage, and C. N. Davies, "Overlapping of Dust Particles on a Sampling Plate," Nature, 163, 809, 1949.

C. E. Lapple, "Particle-Size Analysis and Analyzers," Chem. Engr., 75:11, 149-156, May 20, 1968.

A. Little, "Part C--Size Analysis of Dust," Gas Purification Processes (G. Nonhebel, Ed.), George Newnes, London, 1964.

W. C. McCrone, "Morphological Analysis of Particulate Pollutants," Air Pollution (A. C. Stern, Ed.), Vol. II, Academic Press, New York, 1968, Chapter 22.

W. C. McCrone, R. G. Draftz, and J. G. Delly, The Particle Atlas, Ann Arbor Sci. Publ., Ann Arbor, Michigan, 1967.

Microscopic Analysis of Atmospheric Particulates, Air Pollution Microscopy Training Manual, U.S.P.H.S., Cincinnati, Ohio, January 1963.

K. E. Noll, "A Procedure for Measuring the Size Distribution of Atmospheric Aerosols," The Trend, 19:2, 21-26, April 1967.

J. P. Olivier, G. K. Hickin, and C. Orr, Jr., "Rapid, Automatic Particle Size Analysis in the Subsieve Range," Micromeritics Instrument Corp., Atlanta, Georgia, 17pp.

C. Orr and J. M. Dalla Valle, Fine Particle Measurement: Size, Surface, and Pore Volume, Macmillan, New York, 1960.

C. M. Peterson and H. J. Paulus, "Continuous Monitoring of Aerosols over the 0.001- to 10-Micron Spectrum," Am. Industrial Hyg. Assoc. J., 29:2, 111-122, March-April 1968.

G. Pfefferkorn and R. Blaschke, "Dust Analysis with the Aid of the Scanning Electron Microscope 'Stereoscan'," Staub, 27:7, 30-33, July 1967.

P. C. Reist and W. A. Burgess, "A Comparative Evaluation of Three Aerosol Sensing Methods," Am. Industrial Hyg. Assoc. J., 29:2, 123-128, March-April 1968.

F. M. Renshaw, J. M. Bachman, and J. O. Pierce, "The Use of Midget Impingers and Membrane Filters for Determining Particle Counts," Am. Industrial Hyg. Assoc. J., 30:2, 113-116, March-April 1969.

M. D. Silverman and W. E. Browning, Jr., "Fibrous Filters as Particle-Size Analyzers," Science, 143:3606, 572, 573, February 7, 1964.

G. B. Smith and G. V. Downing, Jr., "Objective Method of Data Reduction for Particle Size Analysis by Cumulative Sedimentation Method," Anal. Chem., 42:1, 136-138, January 1970.

"Standard Method for Particle Size Determination," Dust Topics, 4:3, 12-14, August 1967.

W. Stöber and H. Flachsbart, "Size-Separating Precipitation of Aerosols in a Spinning Spiral Duct," Environmental Sci. Tech., 3:12, 1280-1296, December 1969.

L. O. Sundelöf, "On the Accurate Calculation of Particle-Size Distributions in Aerosols from Impaction Data," Staub, 27:8, 22-28, August 1967.

K. T. Whitby, Determination of Particle Size Distribution--Apparatus and Techniques for Flour Mill Dust, Bull. No. 32, University of Minn. Engr. Expt. Sta., January 1950.

K. T. Whitby and W. E. Clark, "Electric Aerosol Particle Counting and Size Distribution Measuring System for the 0.015 to 1 u Size Range," Tellus, XVIII:2, 573-586, 1966.

Chapter 9

SPECIFIC TESTS

I. INTRODUCTION

Several tests that are more or less accepted for air pollutant evaluation are briefly treated here. The test methods selected are those that are most popular and believed to be the best available. The sensitivity, precision, and accuracy of the method are given when such information is known.

II. PHYSICAL TESTS

Quantification of the physical amounts of some pollution parameters without specific identification testing of the material involved is often desirable. This is especially true for particulates and water vapor.

A. Dustfall

There is disagreement on the correct method for determining dustfall. The method described here is probably the most widely used.

Equipment:

Quart jar (or jar marketed for this purpose or similar; the home canning type of glass jar works very well and is quite economical.), $103^{\circ}C$ oven, desiccator, analytical balance, frame and platform.

Procedure:

1. Determine the tare weight of the quart jar or equivalent, or plan to transfer the collected material to an evaporating dish.

2. Add 1/3-1/2 qt of distilled water. Some workers prefer to use the jar dry.

3. Mount the jar in a rack to prevent its being tipped over by the wind. The rack should be placed on a platform or in the center of a flat roof of a building. The cone of revolution $\geq 45°$ from the vertical should not intersect any obstruction.

4. Keep some water in the jar throughout the test, preferably 1 month or more.

5. Determine the equilibrated dry weight of the sample (Chapter 7, Section IV, E).

6. Calculate the dustfall in tons per square mile per month from the area of the mouth of the jar and the weight of sample collected.

EXAMPLE 9-1: If a jar with a neck diameter of 2 7/16 in. collects 0.048 gm of dust in 45 days, what is the dustfall in tons/mi^2-month?

$$\frac{0.048 \text{ gm}}{9.07 \times 10^5 \text{ gm/ton}} \times \frac{1}{45/30 \text{ months}} \times \frac{\pi}{4} \frac{(39/192 \text{ ft})^2}{(5280 \text{ ft/mi})^2}$$

$$= 30 \text{ tons/mi}^2\text{-month}$$

When a balance with the required sensitivity (milligrams) and capacity (a few hundred grams) is not available, transfer the water and collected dust to a smaller, tared container for evaporation. Use a scraper or policeman and rinse with distilled water to assure that the last vestige of material is transferred.

In freezing weather a liquid other than distilled water should be used. Any liquid that leaves no residue upon evaporating and does not freeze, such as alcohol, may be substituted. A known or negligible weight of algacide or fungicide may be needed to prevent growths. It is advisable to put out multiple jars to insure collection of at least one sample free of dead insects and bird droppings.

Many other methods of measuring dustfall or sootfall have been devised. Jacobs described several methods at length, and reference is made to his listing in the bibliography for those interested in examining the subject further.

Dustfall values may range from about 10 to more than a 1000 $tons/mi^2$-month; usual values for a city are 30-100 $tons/mi^2$-month. Nader made a study with plastic containers and found no significant difference between wet and dry operation. The weights collected in an experiment were about 115 mg and showed standard deviations of 15% of the means. Baffles in the jars, which divided them into four chambers, reduced the standard deviation to about 5% of the mean.

B. Suspended Particulates

The concentration of suspended particulates is used as a measure of the quality of the ambient air and as a control index for some emission regulations.

Equipment:

High-volume sampler with metering device, filter holder (8 X 10 in.), glass fiber filter (8 X 10 in., type A or E), analytical balance rigged to weigh the filter.

Procedure:

1. Filter a large volume of air (1-2 m^3/min for 2-24 hr) through an equilibrated (Chapter 7, Section IV, E), tared 8 X 10 in. fiber glass filter. The sample should be collected at a location and in a manner that are consistent with the purpose of the sampling, weather conditions, and avoidance of resuspended dusts.

2. Determine the weight gain by equilibrated weighing of the filter.

3. Calculate the concentration of suspended particulates in micrograms per cubic meter $(\mu g/m^3)$.

The usual range of suspended matter is from 20 $\mu g/m^3$ in clean mountain air to many thousands of micrograms per cubic meter downwind of a dirty process. The values measured for an entire city or over a wide metropolitan area are likely to follow a log normal distribution or at least a positively skewed

distribution which may be represented by the log normal. When
sampling a concentration of 39 $\mu g/m^3$, the standard deviation
was found to be 6 $\mu g/m^3$ (about 15%) and at 112 $\mu g/m^3$, 10 $\mu g/m^3$
(about 9%).

This method of collection for airborne material does not
determine the total suspended material in the strictest sense
but does so within the usual definition. Some particles evaporate
or sublime, and some very small ones pass through the filter.
The sampler is often placed under a small rain shelter for lengthy
sampling periods to prevent washing of the filtered material in
case of rain; however, fog particles may wash the filter even
with the shelter. The shelter results in lower readings close to
the source or the ground because it misses larger particles which
would be collected without the shelter. This is probably desirable
as the large particles would be rapidly removed and really contribute little to air pollution but considerable to the weight of
the sample. A very detailed description of the shelter and of
the whole sampling procedure is to be found in the Federal
Register reference.

The cost of special purpose equipment for measuring suspended particulates is about $250 for a high-volume sampler
and filter holder and $25 per 100 for the filters. A special weigh-
ing frame is needed for most balances to prevent breakage of the
filter since they will not bend without breaking. For single-pan
balances, the frame may be bought with an enclosing box for
$230, or a saddle may be rigged with wire and ingenuity which
will also serve.

C. Dirtiness Parameters

Dirtiness parameters have been defined to describe the
overall soiling capability of the air and to give a gross indication
of the degree of air pollution. Looking at your shirt collar at the
end of the day indicates the condition very well but does not

allow quantification. Atmospheric visibility is a related measurement which is described in Chapter 10. The soiling index, which
is measured by the dirtying of a filter, is the most common dirtiness measure. A combination of soiling index and sulfur dioxide
has been suggested as an index of air pollution, but this would
be applicable only in coal-burning regions, or areas with sulfur
dioxide problems.

Equipment:

Whatman No. 41 filter paper (tape), paper tape sampler
(filter holder, pump, meter, clock), analyzer photometer
(or light meter).

Procedure:

1. Filter a known volume of air, at 8-15 lpm, through a
spot on a Whatman No. 41 filter (usually a tape).

2. Measure the attenuation of light passing through the
dirty spot (or reflection from the spot). Set 100% transmission
(or reflection) on the clean adjacent filter area. Read percent
transmission (or reflection) through (or from) the dirty spot.

3. Calculate the dirtiness parameter.

The soiling index is calculated from

$$S.I. = \frac{100\,A}{L/1000} = 10^5\,A/L, \qquad (9-1)$$

where S.I. = soiling index (Cohs), A = absorbance (optical
density) = $\log_{10} I_o/I = \log_{10} 100/$percent transmission ($I_o$ is
the incident light and I is the transmitted light), and L = the
length of the sample (ft) = volume sampled (ft^3)/area of spot (ft^2).
The soiling index is often expressed in Cohs/1000 ft (Coh is an
acronym for coefficient of haze, as defined by Hemeon). The
soiling index ranges from a few tenths to 10 Cohs/1000 ft. A
value of a few Cohs indicates very dirty air.

Gruber suggested that reflectance from the dirty spot would
be a better measure of soiling index and defined Ruds (reflectance

units of dirt shade). This soiling index is calculated by

$$S.I. = \frac{E}{L/10,000} ,$$ (9-2)

where S.I. = soiling index (Ruds), E = percent extinction by
dirty spot, and L = length of sample (ft). It is noted that Ruds
are measured per 10,000 ft of sample. The values found are
presumably of the same order of magnitude as for Cohs/1000 ft.
Ruds are used much less than Cohs. They both suffer the same
principal fault; that is, oily or greasy substances may give
negative S.I. values, while causing quite a blackening of the
filter spot.

EXAMPLE 9-2: A 4-hr sample for soiling index shows 40% transmission through the dirty spot. If the sampling rate was
0.35 cfm and the spot 1 in. in diameter, what soiling
index should be reported?

$A = \log_{10} 100/\% \, T = \log_{10} 100/40 = 0.397$
$L = (4 \times 60 \times 0.35 \, \text{ft}^3)/(\pi/4 \times \overline{1/12}^2 \, \text{ft}^2) = 15,500 \, \text{ft}$
$S.I. \, (\text{Cohs}) = 10^5 A/L = 10^5 \times 0.397/15,500 = 2.56$

Most soiling index measurements are made with paper tape
samplers with sampling times adjustable up to 4 hr; the filter is
then moved to a clean spot and the sampling repeated. There is
a pause between sampling periods that may be adjusted from
about 0 to 4 hr. The tape is analyzed by a photometer which is
often built into the tape sampler. The paper tape sampler may
be used for hydrogen sulfide and suspended particulates may be
correlated with the β-radiation transmission through the dirty
spot. The β-radiation attenuation is a much better measure of
the weight of material in the air, but darkness of color has no
effect on the attenuation. Perhaps a combination of visual color
of spot with β-transmission would be the best measure of dirtiness.
Since the analysis is nondestructive, the spot can be used for
running chemical analyses on the suspended particulates. The
cost of the paper tape sampler and the analyzer is approximately
$500.

D. Dew Point

The moisture content of air or stack gas is often measured by passing a filtered volume through silica gel or a condensation trap and noting the weight gain. The weight of water can be used with the temperature to calculate a dew point, or steam tables can be used to obtain the weight of water for saturation and the dew point determined from nomographs. This is not a good way to measure dew point because sulfur trioxide or other gas may change the actual dew point radically. Catalytic hygrometers have been used to measure the moisture content of blast furnace gases and other hot gas streams. The wet bulb and dry bulb temperatures can be used to get the dew point, but this method also suffers from the presence of acidic gases.

The proper measure of dew point for air pollution work is the determination of the actual temperature of a gas stream that starts to produce condensed moisture. Thermocouples may be used to measure the temperature of two surfaces cooled by circulating Freon or other gas while an electric potential is applied across the surfaces; condensation between the surfaces changes the conductivity and permits the dew point to be read. Unfortunately, such instrumentation is not generally available, but it has been described in the literature.

At least one manufacturer makes a dew point instrument that measures the dispersive change in reflectance of a light beam from a bright surface at the temperature of condensation. This instrument is designed for temperatures to $150^{\circ}F$ and cannot be used with most stack gases.

III. CHEMICAL TESTS

Chemical tests for the most important pollutants (sulfur dioxide, nitrogen oxides, ozone and oxidants, aldehydes, and fluorides) are described in sufficient detail to permit running the analyses with these instructions. Some other chemical tests are named with very brief comment (hydrogen sulfide, sulfur trioxide, sulfuric acid, and ammonia). The sampler should employ a trap after the absorber (bubbler) to protect the equipment following. This is usually a mist trap of fiber glass, carborundum, and so on. If a check on the efficiency of the bubbler for collecting the pollutant is desired, a second absorber should be put in series with the first.

A. Sulfur Dioxide

The best method currently available for sulfur dioxide analysis is the West-Gaeke method, or a modification thereof. This method can measure concentrations over an approximate range of 0.005-5 ppm$_V$ with an accuracy of ± 10% (including sampling and analysis) at the lower end of the range and ± 5% at the upper end, with a precision of about 2%.

Ozone and nitrogen dioxide interfere if their concentrations exceed that of the sulfur dioxide. The nitrogen dioxide interference is removed by 0.06% sulfamic acid in the absorbing reagent, but the sulfamic acid may cause loss of sulfur dioxide during protracted sampling or storage of reagent.

The sampling should be made with probes and lead-ins of Pyrex glass, stainless steel, or Teflon. Butt-to-butt glass tubing may be connected with Tygon. Never allow the sample to contact rubber.

Equipment:

Midget impinger, trap, meter, pump, spectrophotometer or colorimeter (560 nm). In the absence of a spectrophotometer or colorimeter, visual comparison can be made in tubes. The light paths should be long and of the same length. A set of Nessler tubes is preferred.

Preparation of Chemicals:

Absorbent: 0.1 M sodium tetrachloromercurate (Na_2HgCl_4).

Dissolve 27.2 gm mercuric chloride and 11.7 gm sodium chloride in 1 liter of distilled water. (Caution: Poisonous! Flush off skin immediately.)

Coloring agent: 0.04% pararosaniline hydrochloride. Dissolve 0.2 gm of reagent in 100 ml of distilled water; let stand 48 hr and filter. Solution is stable at least 3 months in a dark, cool place. Put 20 ml of solution in 100-ml flask, add 6 ml of concentrated hydrochloric acid and, after 5 min, fill to mark with distilled water. Solution is stable 1 week in amber bottle or 2 weeks if refrigerated.

0.2% Formaldehyde. Dilute 5 ml of 40% formaldehyde to 1000 ml with distilled water. Solution is stable 1 week.

Calibration:

0.0123 N Sodium metabisulfite (1 ml = 150 μl of sulfur dioxide). Dissolve 640 mg of reagent (65.5% as sulfur dioxide) in 1 liter of distilled water. Standardize to 0.0123 N by titration with 0.01 N iodine, using starch as an indicator.

0.01 N Iodine. Dissolve 12.69 gm of resublimed iodine in 25 ml of solution made with 15 gm iodate-free potassium iodide (KI), dilute to 1 liter, pipet 100 ml into 1000 ml flask, fill to mark with 1.5% potassium iodide. Used as a standard or checked by standard thiosulfate.

Procedure:

1. Put 10 ml of absorbent in midget impinger (all glass).

2. Bubble enough sample through to obtain 2-4 μl of sulfur dioxide. Sampling rate of 0.1 cfm (2.8 lpm) or less. Dichloro-sulfito-mercurate ion is formed. Solution may be stored up to 3 days without correction.

3. Filter if particles are present. Adjust volume to 10 ml with distilled water.

4. Add 1 ml each of complexing reagents, pararosaniline hydrochloride and formaldehyde, and mix.

5. Prepare a 10-ml absorbent blank (unexposed) in same manner.

6. After 20 min, read absorbance at 560 nm on spectrophotometer or colorimeter using blank as a reference.

7. Use calibration curve for microliters of sulfur dioxide.

8. Calculate parts per million (or micrograms per cubic meter) of sulfur dioxide. 1 ppm_V = 1 μl of sulfur dioxide per liter of air sampled.

Calibration Curve:

The light transmission should follow Beer's law (Chapter 7, Section III, A, 2). The slope should be about 0.15 absorbance units per microliter of sulfur dioxide with a 1-cm light path or a line through 100% transmittance at 0 concentration of sulfur dioxide and about 71% at 1 μl of sulfur dioxide.

Preparation:

1. Pipet 2.00 ml of 0.0123 N sodium metabisulfite into a 100-ml volumetric flask. Fill to mark with absorbent. 1 ml = 3.0 μl of sulfur dioxide.

2. Add 0.00, 0.50, 1.00, 1.50, and 2.00 ml to each of five 10-ml glass stoppered cylinders.

3. Develop color as in steps 4-6 above.

4. Plot absorbance versus microliters of sulfur dioxide on arithmetic graph paper or percent transmittance (log) versus microliters of sulfur dioxide (arith) on semilog paper.

B. Nitrogen Dioxide and Nitric Oxide

The Saltzman method is probably the best method for measuring nitrogen dioxide and nitric oxide. The method is applicable over a range of about 0.005-5 ppm_V. Errors of sampling and interference are small. The small interference by high concentrations of ozone and sulfur dioxide can be eliminated by reading the color in 45 min, or for the sulfur dioxide by adding 1% acetone to the reagent before use.

Equipment:

Special absorber (see Fig. 9-1) or midget impinger (with
calibration correction), trap, meter, pump, spectrophotometer
or colorimeter (550 nm), midget impinger bubbler for nitric oxide.

Preparation of Chemicals:

Absorbing reagent: Dissolve 5 gm sulfanilic acid in a
1000-ml volumetric flask with a mixture of about 800 ml of
distilled water and 140 ml of glacial acetic acid by gentle
heating. Cool the mixture and add 20 ml of 0.1% n-(1-napthyl)-
ethylenediamine dihydrochloride (EDDHC). Fill to mark with
distilled water. 0.1% EDDHC: Dissolve 0.1 gm of the reagent
in 100 ml of distilled water.

Standard: 1 ml = 10 μl nitrogen dioxide. Prepare a stock
solution by dissolving 2.03 gm sodium nitrite to 1 liter with
distilled water. Prepare a fresh calibration standard when ready
to use by putting 1 ml of the stock in a 100-ml flask and filling
to mark with distilled water. 0.0202 gm/liter.

Nitric oxide reagent: Use a 100-ml volumetric container
to dissolve 2.5 gm potassium permanganate in about 90 ml of

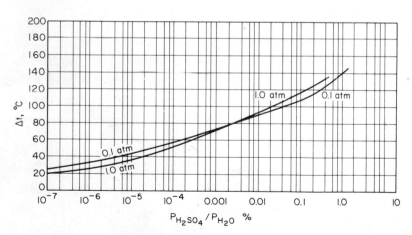

FIG. 9-1. Sulfur trioxide elevation of dewpoint. From
P. Müller, Chemie Ing. Technik, 31, 345, 1959.

distilled water. Add 2.5 gm (about 1.75 ml) of concentrated
sulfuric acid. Fill to mark with water. When brown precipitate
is considerable, discard and make a fresh solution.

Procedure:

1. Pipet 10.00 ml of absorbent into special absorber.

2. Bubble air through at ≤ 0.4 liters/min until sufficient
color has developed, about 10 min. Record volume of air sampled.

3. Allow 15 min for color development.

4. Read absorbance (optical density) or percent transmittance at 550 nm.

5. Use calibration curve to obtain microliters of nitrogen
dioxide.

6. Calculate parts per million of nitrogen dioxide.
1 ppm_V = 1 μl nitrogen dioxide per liter of sampled air.

Calibration Curve:

1. Add 0.00, 0.25, 0.50, 0.75, and 1.00 ml of standard
sodium nitrite solution to each of five 25-ml volumetric containers.

2. Dilute to marks with absorbing reagent.

3. After 15 min, read color at 550 nm.

4. Plot calibration curve (as for sulfur dioxide).

For nitric oxide determination, follow the nitrogen dioxide
absorber with an acid permanganate bubbler to convert nitric oxide
to nitrogen dioxide, then absorb the nitrogen dioxide thus formed
in a second special absorber. If nitric oxide is desired without
nitrogen dioxide, simply start the sampling train with the conversion unit.

C. Ozone and Oxidant

The neutral buffered potassium iodide method is recommended
by Saltzman as the best method for the determination of oxidants
(ozone and others) despite negative interferences from reducing
substances such as hydrogen sulfide, sulfur dioxide, and others.
The test can measure oxidants (such as ozone) in a range from a

few hundredths to about 10 ppm$_v$. The size of the errors in the test are dependent upon the amount of interfering substances. The analysis must be completed during 30 min to 1 hr after sampling because the last 10% of the color reaction has a half-life of about 10 min. The color increases to about 45 min then decreases.

The sulfur dioxide interference can be eliminated by pulling the sample through a U-tube (140 ml) packed with chromium trioxide impregnated paper strips ($\frac{1}{4}$ X $\frac{1}{2}$ in.). Drop 15 ml of aqueous solution with 2.5 gm chromium trioxide and 0.7 ml sulfuric acid uniformly over 60 in^2 of paper and dry in an oven at 80-90°C for 1 hr. Cut into strips, fold into V-shape, and pack tube. Condition the tube by drawing air through it overnight. The paper lasts 1 month and may be dried if it becomes wet.

Equipment:

Midget impingers (all glass), trap, meter, pump, spectrophotometer or colorimeter (352 nm).

Preparation of Chemicals:

Solution water: To be used for all reagents. Double-distilled water, second distillation in all-glass still with a crystal of potassium permanganate ($KMnO_4$) and barium hydroxide [$Ba(OH)_2$].

Absorbent: Dissolve 13.61 gm of potassium dihydrogen phosphate, 14.20 gm of anhydrous disodium hydrogen phosphate (or 35.82 gm of dodecahydrate salt), and 10.00 gm of potassium iodide, one at a time, and dilute to 1 liter. Age at least 1 day at room temperature. The solution has a lifetime of several weeks in brown bottle away from sunlight.

Standard solution: 0.05 N iodine solution. Dissolve (in order) 16.0 gm of potassium iodide and 3.173 gm of iodine and dilute to 500 ml. Age at least 1 day before using.

Procedure:

Use only glass upstream of absorber. Do not expose absorbent to direct sunlight.

1. Pipet 10 ml of absorbent into midget impinger.

2. Sample at a flow rate of 1-2 liters /min for up to 30 min. (0.5-10 µl of ozone or equivalent should be collected.)

3. Add distilled water to 10-ml mark.

4. Measure absorbance or percent transmittance at 352 nm during period 30-60 min after sampling.

5. Subtract blank correction. A calibration check should be run at least every few days.

6. Use calibration curve to obtain microliters of oxidant (as ozone). 1 ppm$_v$ = 1 µl oxidant/liter of sampled air.

Calibration:

1. Prepare 0.0025 N iodine from 5 ml of 0.05 N stock solution in 100-ml flask and fill to mark.

2. Pipet 0.00, 0.20, 0.40, 0.60, and 0.90 ml into each of five 25-ml volumetric containers.

3. Fill to marks with absorbent.

4. Immediately read colors at 352 nm.

5. Plot curve of absorbance versus microliters of ozone (0.00, 2.448, 4.896, 7.344, and 11.016 for standards; calculated by 12.24 X ml of standard). The curve should go through 4.8 microliters of ozone at absorbance of 1 for 2-cm cells.

D. Fluorides

It is often desirable to measure the particulate and the gaseous fluorides separately. This may be accomplished by filtering the particulates before the gaseous fluorides are absorbed. An impregnated filter can be used to collect both the particulate and the gaseous fluorides. The tests that have been used for fluorides have left much to be desired. The current trend is toward the use of ion-specific electrodes.

1. Ion-Specific Electrode Method

The ion-specific electrode is used to measure airborne fluorides in a method reported by Elfers and Decker. The method is claimed to measure levels of fluorides as low as 0.25 ppb$_v$.

Equipment:

F$^-$-Specific electrode (Orion Model 94-09), potassium chloride electrode, pH meter for measuring potential, 2-in. or 47-mm membrane filters (usually 0.8 μm pore size), pump, meter.

Preparation of Chemicals:

Filter impregnation: Prepare a solution with 50% ethanol and 10% sodium formate.

Calibration: Sodium fluoride (F$^-$ = 10 mg/ml). Dissolve 22.105 gm of sodium fluoride in distilled water and fill to 1 liter.

Buffer: 0.1 M sodium citrate

Procedure:

1. Dip 2-in. membrane filters in solution of sodium formate. Dry filters at room temperature.

2. Pull 0.4 cfm of air for 4 hr through the treated filter.

3. Place filter in 25-ml plastic vial with 20 ml of 0.1 M sodium citrate buffer.

4. Shake for 1 hr on mechanical shaker.

5. Transfer solution to 50-ml polyethylene beaker.

6. Measure potential with pH meter (F$^-$-specific electrode and potassium chloride electrode).

7. Use calibration curve to obtain milligrams F$^-$ per liter of liquid.

8. Calculate concentration of F$^-$ in micrograms per cubic meter of air.

Calibration:

1. Dilute F$^-$ standard solutions (0.0, 0.1, 1, 10, 100, and 1000 mg F$^-$/liter by adding 0.00, 0.01, 0.10, 1.00, 10.00, and 100.00 ml to each of five 1-liter flasks. Fill to marks.

2. Buffer each to pH 7.5 with 0.1 M sodium citrate.

3. Measure potential with pH meter using special electrodes.

4. Plot millivolt potential versus F$^-$ concentration in milligrams per liter.

For stack gas fluorides, sample with a train consisting of the following: heated glass probe, cyclone, Whatman No. 41 filter, and a series of Greenburg-Smith impingers. Particulate fluorides are removed in the cyclone and on the filter. Water-soluble gaseous fluorides are collected in the first two impingers. A 10-ml composite from the two impingers is combined with 10 ml of 0.2 M sodium citrate buffer and analyzed by an ion-specific electrode.

2. Wet Chemistry Method

The Willard and Winter distillation procedure is the most often used wet chemistry method for the determination of fluorides, although the Spadns-Zirconium-Lake method is being used by several laboratories.

Equipment:

High-volume filter (fiber glass) or midget impinger, meter, pump, muffle furnace, 150°C oven, platinum crucible, distillation apparatus (see Fig. 9-2).

Procedure:

For suspended particulates

1. Put glass fiber filter (or portion thereof) that has filtered 1500-2000 m^3 of air in 24 hr (less for higher concentrations) in a platinum dish.

2. Cover with 10 ml of calcium hydroxide suspension.

3. Evaporate to dryness on a steam bath.

4. Heat for 30 min in 150°C oven.

5. Ignite in muffle furnace at 550°C for 5-6 hr.

6. Break up ash in the dish with a glass rod.

7. Mix in 1 gm of powdered silver perchlorate.

8. Transfer the ash-silver perchloride mixture to the distillation apparatus with the aid of water and a few drops of 60% perchloric acid rinses.

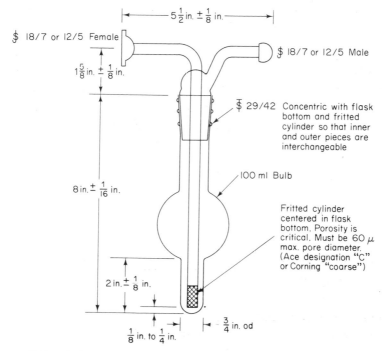

$ 18/7 or 12/5 Female

$ 18/7 or 12/5 Male

$ 29/42 Concentric with flask
bottom and fritted
cylinder so that inner
and outer pieces are
interchangeable

100 ml Bulb

Fritted cylinder
centered in flask
bottom. Porosity is
critical. Must be 60 μ
max. pore diameter.
(Ace designation "C"
or Corning "coarse")

$5\frac{1}{2}$ in. $\pm\frac{1}{8}$ in.

$1\frac{5}{8}$ in. $\pm\frac{1}{8}$ in.

8 in. $\pm\frac{1}{16}$ in.

2 in. $\pm\frac{1}{8}$ in.

$\frac{1}{8}$ in. to $\frac{1}{4}$ in.

$\frac{3}{4}$ in. od

FIG. 9-2. Nitrogen dioxide bubbler. From B. E. Saltzman, Selected Methods for the Measurement of Air Pollutants, U.S. P.H.S., Cincinnati, Ohio, May 1965.

9. Add 10 ml of 60% perchloric acid, two or three small pieces of pure silica, and two small glass beads.

10. Steam distill at 135°C with heat under the flask and steam flow both regulating the temperature (see Fig. 9-3).

11. Collect about 190 ml of distillate and make to 200 ml with water.

12. Use 50 ml of solution for colorimetry. Adjust temperature to that used for standard curve. Add 5.0 ml of zirconyl chloride acid solution and mix well. Transfer to a special cuvette and read absorbance or transmittance at 530 nm. Run blank on the same amount of unexposed glass fiber filter.

13. Express net fluoride as F^- in micrograms per cubic meter.

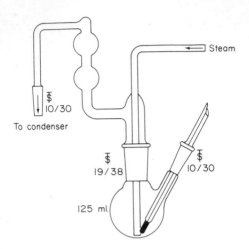

FIG. 9-3. Fluorides distillation apparatus. From M. B.
Jacobs, The Chemical Analysis of Air Pollutants, Interscience,
New York, 1960.

Alternate Procedure:

For soluble gases and particles

1. Bubble air through 0.1 N sodium hydroxide in Greenburg-
Smith impingers (followed by trap). Because of the large volume
needed, use 24 samplers collecting simultaneously for 1 hr or
serially for 24 hr.

2. Transfer to 2-liter beaker.

3. Evaporate to small volume, transfer to small beaker,
and evaporate to small amount.

4. Transfer to a platinum dish with rinses.

5. Evaporate to small volume.

6. Start with step 2 of the above procedure.

E. Aldehydes

The fact that formaldehyde constitutes about one-half of the
total atmospheric aldehydes is often used in the determination
of formaldehyde from measurements of total aldehydes or vice
versa. The MBTH method is used to measure total aldehydes by

a colorimetric procedure or, at other wavelengths (635 or 670 nm), formaldehyde.

Equipment:

Bubbler ($\geq$ 35 ml capacity), trap, meter, pump, spectrophotometer or colorimeter (628 nm).

Chemicals:

Absorbing: 0.05% Aqueous solution of 3-methyl-2-benzothiazolone hydrazone (Aldrich or Eastman).

Oxidizing: 1.6% Sulfamic acid and 1% ferric chloride in aqueous solution.

Calibrating: 1 mg formaldehyde/ml. Dilute 2.7 ml of 37-39% (commercial reagent) formaldehyde to 1 liter. (Note: Solution may be standardized by dimedon; see bibliographical item Selected Methods for the Measurement of Air Pollutants.)

Procedure:

1. Bubble 0.5 liters/min of air through 35 ml of absorber for 24 hr.

2. Restore to 35 ml with distilled water.

3. Put 10 ml of sample into test tube.

4. Add 2 ml of oxidizer, mix, and let stand for 12 min.

5. Measure absorbance at 628 nm against blank of 10-ml absorber and 2 ml of oxidizer.

6. Calculate parts per million of total aldehydes (as formaldehyde).

$$ppm_v = 0.814 \text{ (35 ml) X (micrograms of aldehyde)}$$

Calibration:

1. Make five test solutions containing 0.1-0.7 µg of formaldehyde per milliliter of absorber.

2. Repeat steps 3 to 5 above.

3. Plot absorbance versus micrograms of formaldehyde per milliliter of absorber.

F. Other Test Methods

Some other important analytical methods are outlined in brief. The bibliography lists references containing more detailed information on these methods.

1. Nitrates

The test described for nitrates is the one used by the NASN. The nitrates are leached (refluxed with water) from a filter or portion of filter containing the particulates from at least 20 m^3 of air; the leachate (5 ml + 15 ml 85% sulfuric acid) is used to nitrate 1 ml of 2,4-xylenol reagent (5 ml:500 ml glacial acetic acid), which is extracted with 10 ml of toluene, separated from the water, and 10 ml of 0.4 N sodium hydroxide added; the color is measured at 435 nm.

2. Sulfates

Nephelometry is used to measure barium sulfate turbidity as an indication of the sulfates. A water extract (20 ml of 50 ml total) from the filter (as for nitrous oxide) is placed in tube; add 1 ml of 10 N hydrochloric acid; add 4 ml of glycerol-alcohol (1:2); add about 0.25 gm barium chloride crystals; let stand 40 min; read absorbance at 500 nm versus blank without crystals added.

3. Hydrogen Sulfide

The methylene blue method is the method of choice for the determination of hydrogen sulfide in the air. Bubble 0.1 cfm of air through 10 ml of absorber (4.3 gm $CdSO_4 \cdot 8\ H_2O$ + 0.3 gm sodium hydroxide to 1 liter); add 0.6 ml of amine-acid (30 ml water to 5 ml sulfuric acid, cool, add and dissolve 12 gm N,N-dimethyl-p-phenylenediamine, make 25 ml to 1 liter with 1:1 sulfuric acid); add 1 drop of ferric chloride solution (100 gm $FeCl_3 \cdot 6\ H_2O$ to 100 ml); let stand 30 min; read color at 670 nm.

Lead acetate can be used to impregnate paper tape (or porous tile) for direct reaction with air passing through. The amount of discoloration is the measure of hydrogen sulfide.

4. Polynuclear Hydrocarbons

Piperonal chloride gives color reactions with polynuclear hydrocarbons. The piperonal chloride is prepared by mixing 21 gm phosphorus pentachloride and 15 gm piperonal to liquid; warm to 45°C for 2-4 min, cool, add excess ice water, and filter the precipitate. Dissolve a portion of the residue from the benzene-soluble test (see the following section) in a small amount of chloroform; add 5 ml of trifluoracetic acid, then a few crystals of piperonal chloride; measure the color produced: (benzo [a] pyrene at 753 and 563 nm; anthracene at 732 nm; naphthalene at 590 nm; and many other substances at 400-750 nm).

5. Organics

There is no exact way to determine the organics in particles; however, some approximate procedures have been defined. The principal such method is the measurement of the benzene-soluble material. A one-third portion of the 8 X 10 in. fiber glass filter used for the suspended particulate measurement is folded and tied with copper wire, then placed in a Soxhlet extraction apparatus and refluxed for 6-8 hr. The weight of the residue after the benzene has been distilled off gives a measure of the benzene-soluble fraction.

Other methods of organics measurement include the use of a solvent other than benzene (e.g., acetone) and the loss of weight when the filtered material is burned at 600°C for 30 min.

6. Inorganics

The portion of the particulates that is insoluble in benzene or remains after burning is termed the inorganic fraction.

IV. BIOLOGICAL TESTS

It is often desirable to measure the biological materials found in ambient air. These measurements are usually made for bacteria or their spores, algae, and pollen.

A. Bacterial Count

Airborne bacteria are normally quantified in a gross manner by counting all of the bacteria that grow on nutrient agar
and using the count as a measure of the concentration of microorganisms. The bacteria may be collected directly on the nutrient
agar or they may be collected in a liquid, then transferred to the
agar plates.

Equipment:

Andersen sampler or all-glass impingers (Millipore),
nutrient agar, petri dishes, autoclave, incubator.

Procedure:

For Andersen Sampler

1. Prepare six sterile petri dishes with nutrient agar.
The dishes are sterilized by 15 psi of live steam in an autoclave.
The agar is prepared according to the directions on the package
and sterilized in the autoclave. Enough agar is applied to each
plate to cover the bottom--be careful to avoid contamination of
the plates.

2. Sample at 1 cfm in the desired location for a time
sufficient to collect a statistically significant number of bacteria
but not overwhelm the plates (usually from a fraction of a minute
to a few minutes). It is desirable to have 100 to 200 colonies
on the plate with the maximum number.

3. Cover the plates, invert, and place in a $20^{\circ}C$ incubator (room temperature is all right) for 48 hr.

4. Count the number of colonies present. If the number
is small, a direct count of each plate will suffice; however,
for large numbers, a single colony may include more than one
bacterial cell from the atmosphere and a correction must be made.
There are 400 holes per stage and the probable number of bacteria
sampled on a stage may be calculated by

$$N = 400 \left[1/400 + 1/399 + \ldots + 1/(400 - n + 1) \right], \quad (9\text{-}3)$$

where N = probable number of bacteria collected and n = number
of colonies observed. Tables are available for solving this
equation (see Andersen).

5. Record the various sizes according to plate number
if sizing is desired. The first two stages collect those particles
greater than 5 μm and the last four stages those less than 5 μm.

6. Calculate the bacterial concentration as number per
cubic meter or per cubic foot.

For all-glass impinger

1. Put 30 ml of 2% heart-brain infusion in the impinger.
Sterilize the impinger and contents in the autoclave as above.
This mixture is used in an effort to sustain the bacteria without
an increase in number.

2. Sample the air at the desired location with a 1-cfm
rate for a few minutes.

3. Filter portions of the sampling fluid through sterile
membrane filters, and place the filters on nutrient broth-saturated
blotter pads in petri dishes. By use of three different size portions (separated by factors of 10x), success in obtaining a
manageable number of colonies is insured.

4. Incubate for 48 hr at $20^{\circ}C$.

5. Count the number of colonies and calculate the
bacterial density in the atmosphere. Each colony is assumed
to have been caused by one bacterial cell and no correction is
made.

For both procedures it is advisable to inspect the plates
at 12 and 24 hr to see if they should be counted in 24 hr to avoid
a running together of the colonies which might occur in the longer
incubation period.

B. Other Methods

Air sampling for other life forms may be carried out in a
manner similar to the bacterial count with substitution of the

proper growth medium. Algal agars are available to culture algae on plates for the purpose of making plate counts. The incubation periods may be several days to 2 or 3 weeks. Spores can be sampled and differentiated from vegetative cells by use of their greater resistance to environmental stresses such as heat, chemicals, dessication, and lack of food. Different groups within the bacteria may be assayed with the use of differential media; for example, EMB agar for coliforms. Fungi and their spores can be counted for some applications, especially in allergen studies.

Pollen sampling and counting may be carried out in the same manner as for any other airborne particles; however, the usual samplers are rotating rods which impact the particles or rotating slides which centrifugally remove the particles. These types of samplers are commercially available. The identification is made by microscopic viewing. The pollen particles are large (>about 20 μm) and easily identifiable by their shapes, especially with scanning electron microscopes.

V. SUMMARY

There are many analytical methods now available, usually several for any one substance. The methods given here are the most popular methods at the present time, and are the ones that require the least sophisticated analytical equipment. They are generally the ones used to check the specific instruments for measurement of the various pollutants. They are certain to change with time.

A number of physical parameters are used to characterize air pollution. Such indexes deal mostly with dirtiness; they include dustfall, suspended particulates, soiling index, and visibility. All of these measurements are described here, except visibility which is treated in Chapter 10. In addition to the tests for air pollution, there are the physical tests for the identification of pollutants or to aid in their collection which were described in Chapters 7 and 8 and dew point, which is described here.

The analytical chemical methods given here for sulfur dioxide, nitrogen dioxide and nitric oxide, ozone and oxidant, fluorides, aldehydes, and other pollutants were abstracted from the literature listed in the bibliography. These sources should be consulted for clarification of any confusion in the application of any chemical tests.

Biological tests may be carried out to measure the number of viable microorganisms per volume of air or to determine the amount of biological material, usually pollen, in the air. Methods for making bacterial counts are described in full. Brief reference is made to other methods of measurement.

PROBLEMS

1. What are the probable error limits on the 30 tons/mi^2-month in Example 9-1 if the weight can be off 3 in the last digit and the diameter of the jar can be wrong up to 1/32 in.?

2. How long should suspended particulate sampling be carried on at 1.6 m^3/min if the atmospheric concentration is estimated at 200 $\mu g/m^3$ and it is desired to collect at least 30 mg?

3. What is the dew point of a gas stream which has 55%$_v$ of water vapor at 550°F and 50 ppm_v of sulfur trioxide?$_v$

4. What is the soiling index in Cohs when a 2-hr sample of 10 liters/min through a 9/16-in. diameter spot gives an absorbance (optical density) of 0.7?

5. Verify that 2.03 gm/liter of sodium nitrite gives a solution with 1 ml equivalent to 10 μl of nitrogen dioxide.

6. Verify the 12.24 factor calibration of the ozone (oxidant) test.

BIBLIOGRAPHY

Analysis of Atmospheric Inorganics, Training Course Manual, U.S.P.H.S., Cincinnati, Ohio, March 1963.

Analysis of Atmospheric Organics, Training Course Manual, U.S.P.H.S., Cincinnati, Ohio, April 1963.

A. A. Andersen, "New Sampler for the Collection, Sizing and Enumeration of Viable Airborne Particles," J. Bacteriol., 76:11, 471-484, November 1958.

L. A. Elfers and C. E. Decker, "Determination of Fluoride in
Air and Stack Gas Samples by Use of an Ion Specific
Electrode," Anal. Chem., 40:11, 1658-1661, September
1968.

M. B. Jacobs, The Chemical Analysis of Air Pollutants, Interscience, New York, 1960.

M. Katz, "Analysis of Inorganic Gaseous Pollutants," and A. P.
Altshuller, "Analysis of Organic Gaseous Pollutants," and
P. W. West, "Chemical Analysis of Inorganic Particulate
Pollutants," and D. Hoffmann and E. L. Wynder, "Chemi-
cal Analysis and Carcinogenic Bioassays of Organic Par-
ticulate Pollutants," Air Pollution, Vol. II: Analysis,
Monitoring, and Surveying, 2nd Ed.,(A. C. Stern, Ed.),
Academic Press, New York, 1968.

"Procedure for Determination of Suspended Particulates (High
Volume Method)," Federal Register, 36:21, 1507-1509,
January 30, 1971.

B. E. Saltzman, "Standardization of Methods for Measurement
of Air Pollutants," J. Air Pollution Control Assoc., 18:5,
326-328, May 1968.

Sampling and Identification of Aero-Allergens, Training Course
Manual, U.S.P.H.S., Cincinnati, Ohio, September 1962.

Selected Methods for the Measurement of Air Pollutants, U.S.
P.H.S. Publ. No. 999-AP-11, Cincinnati, Ohio, May 1965.
(S. Hocheiser, West and Gaeke Method; B. E. Saltzman,
Nitrogen Dioxide and Nitric Oxide, Ozone and Oxidants;
T. R. Hauser, MBTH Method for Aldehydes).

P. W. West and G. C. Gaeke, "Fixation of Sulfur Dioxide as
Disulfitomercurate (II) and Subsequent Colorimetric Estimation," Anal. Chem. 28, 1816-1819, December 1956.

Chapter 10

EFFECTS OF AIR POLLUTION

I. INTRODUCTION

Air pollution causes physical, chemical, and biological
change or, within the definition of air pollution used here, damage. Some of these effects are patently obvious--others can be
found only in the depths of statistical analyses.

In Chapter 1 some of the major dramatic effects were cited,
both the real catastrophes of the past and the speculative catastrophes of the future. Extrapolations of short-term trends in data,
whether real or fancied trends, will continue to be used for pre-
dicting the end of civilization from one cause or another, especially by those seeking attention or financial support. The
coverage here is mostly of the real effects that can be seen
and quantified and does not usually cover sociological, psychological, political, or moral aspects of air pollution. The real
exception to this guideline is in the section on ecology, wherein
such discussion is considered relevant.

II. EFFECTS IN GENERAL

This section describes the broad aspects of air pollution,
the effects from nonspecific air pollutants either singly or in
combination. In addition, mention is made of some pollutional
effects that are not treated further.

A. Physical

Some of the effects of air pollution are physical in nature.
Such effects are attributable to the transmission of or the absorp-

365

tion or reflection of electromagnetic and particulate radiations or sound waves, or to the removal of pollutants from the air by physical means such as sedimentation (or rising), impaction, absorption, and adsorption.

Physical effects may be visual (dirtiness or visibility attenuation), acoustical (noise or blast propagation), or molecular energetic (electrical energy of ionization or kinetic energy of heat).

Dirtiness affects exposed surfaces and is especially noticeable on clothes, buildings, automobiles, and even trees and plants, just as long as the color of the particulate contrasts with that of the surface. The classic griminess from coal burning has been removed from many buildings in London, Pittsburgh, and other cities that have cleaned up some of their worst pollution sources. The portions of some European cities that were replaced after World War II offer a startling contrast to the old buildings remaining--white stone facades of the new versus sooty black for the old. Dirtiness also clogs air filters in autos, air conditioners, air compressors, and so on.

Obscuration or visibility attenuation is frequently taken as a measure of the degree of air pollution. Many of the air pollution regulations being formulated at the present include visibility clauses. Visibility effects are discussed at length in Section III, A, 4.

Noise and blast may cause structural damage from overpressures or from fatigue. Noise generation capabilities have increased markedly in the last several years, with jet and rocket engines being the strongest sources. Blasts have gone from 1 ton TNT to multimegaton thermonuclear devices. Noise and blast are discussed further in Section III.

Weather modification on a worldwide climatological basis has been of much concern to geophysicists. A review of the greenhouse effect by carbon dioxide and speculations about the overall weather effects from air pollution are considered in an article by Peterson. The known amounts of fossil fuels could raise carbon dioxide concentrations to nearly 5 times their present levels. A doubling of the current 330 ppm_v would raise the surface temperatures by about 4.25° F by absorbing and reradiating IR radiation in the 12-18 region. Carbon dioxide stimulates plant growth, but the effects of carbon dioxide on the overall flora of the world are not known. The uptake by the oceans may take 1000 yr, which is much too slow for alleviating the present rate of carbon dioxide increase before reaching the levels predicted to be cataclysmic from the melting of the polar ice caps. The water in the ice caps (about 7.2 million mi^3) would raise the level of the oceans about 200 ft. Carbon dioxide has increased from about 282 ppm_v in 1860 to its present level of about 330 ppm_v. From 1958 to 1962 the increase was 3.7 ppm_v, or 1.13%, which was about half of the carbon dioxide released from the combustion of fossil fuels.

Stratospheric inventories of long-residence particles from volcanoes, nuclear devices, and other sources reduce the global surface temperatures. The vagaries of weather statistics, their fluctuations and limited number of sampling stations, permit the citing of cycles or trends to support the melting of the ice caps or another ice age or both.

EXAMPLE 10-1: How long will it take to double the carbon dioxide concentration of the atmosphere if the current rate of increase continues? What will be the net temperature change if the $\frac{1}{2}^\circ$C decline reported for the last 50 years is real and continues?

Carbon dioxide change: 1.13% in 4 years (1958-1962)

$1.0113^n = 2$ $n = 61.7$ periods or about 250 years

Temperature rise from carbon dioxide increase $= 4\frac{1}{4}°F$

Temperature decline from particulates increase:

$$(250/50) \times \frac{1}{2}°C \times 9/5 = 4\frac{1}{2}°F$$

Net effect $= 4\frac{1}{2}°F - 4\frac{1}{4}°F = -\frac{1}{4}°F$

Such effects would take a long time in man's calendar--hundreds or thousands of years--but a short time in the geological setting. Mostly, it can be concluded that there should be continued monitoring of the various factors with cognizance of possible catastrophic effects; and, with the current attitudes, there is little doubt that such will be done.

Changes in the radiations or ions in our atmosphere may result from upsets in space radiations, especially solar activity, or from radiations from nuclear devices or locally from heat sources, lightning, and other means. These radiations and ions cause disruption of communications that utilize atmospheric transmission, and they can cause chemical and biological damage. This subject is covered in more detail in Section III, J.

Historically, the physical effects of air pollution have probably received more attention than is warranted by consideration of their contribution to total air pollution damage. This has been a result of the ability of everyone to observe stack emissions of dust or smoke and accumulations of the small particles in an inversion layer. With the emphasis on control for esthetic reasons, much attention to the physical aspects will continue.

B. Chemical

The chemical entities in the atmosphere are subject to continuous change, at greatly varying rates, to other forms. Because of thermodynamic considerations, the expected trend is toward the oxidative change to simpler, more stable products with less internal energy than their precursors; however, there is some disruption of this orderly trend by photochemical and other driven reactions that produce complex and highly unstable substances.

Biological reactions are of little importance in the atmosphere;
they occur mostly at interfaces between solids and liquids.

As a result of the generally unchanging nature of the elements
and the stabilities of certain substances, science has placed
emphasis on the cycles of certain elements and molecules in the
environment; for example, the carbon, nitrogen, sulfur, and
hydrologic cycles.

The atmospheric chemical reactions that may occur in
significant amounts on a worldwide basis can be hypothesized
by considering the composition of the air and the energy available
to drive the reactions. Only four gases are present at levels that
justify listing by percentage; namely, nitrogen, oxygen, argon,
and carbon dioxide (see Table 1-1), and argon is so stable as to
be unimportant in atmospheric chemistry. The energies available
at low elevations in large quantities from the solar spectrum have
wavelengths greater than 300 nm (3000 A) or energies less than
4.1 eV; Table 10-1 shows bond energies for the most important

TABLE 10-1

Thermochemical Bond Energies

Bond	Energy		Photon wavelength (nm)
	(eV)	(kcal/mole)	
C--N	2.10	48.6	590
C--C	2.54	58.6	488
C--O	3.03	70.0	409
C--H	3.79	87.3	327
S--H	3.80	87.5	326
C==N	4.10	94.0	302
C==C	4.33	100.0	286
O--H	4.78	110.0	260

bonds. The upper atmosphere receives more energetic photons
and results in the ozone layer at heights of 10-20 mi. The
prevalent reactions on a global scale involve nitrogen and oxygen
because of their mass action effects on reaction rates. Lightning,
ionizing radiations, and intense heat sources may result in free
radicals and drive reactions on a local or regional scope. Photochemistry is treated more fully in Section III.

Oxidations of an air pollution nature that occur at surfaces
are costly. These are made up on corrosion, combustion, and
explosion. The last-mentioned two need little if any elucidation.
Corrosion is enhanced by many of the common pollutants found in
the atmosphere, but especially by the acid-forming gases and
particles and the acid mists. Corrosion occurs only slowly in
arctic and desert regions and is most rapid in areas of industrial
or salt spray pollution. Chemical erosion, which is also
frequently called corrosion, is important in some instances; for
example, acids on concrete and stone structures, hydrogen
fluoride on glass, acid-forming particles on paints, and similar
surface reactions.

C. Biological

There is widespread disagreement among those knowledgeable
in air pollution concerning the biological effects from ambient
levels of air pollution. The concrete information available on the
chronic effects of relatively low concentrations of air pollutants
has been derived by the statistical methods of epidemiology.
Despite the high probabilibies of cause-effect relations where high
pollution levels are accompanied by high morbidity and mortality
from respiratory diseases, many people remain unconvinced that
the cause can be separated from the complex mixture of exposure
to many health insults that are prevalent in the test population.

The alternative to using epidemiological investigation is
toxicological experimentation. The experiments on animals are

usually run with rather large amounts of the test material, often administered by means other than inhalation, and the results extrapolated to lower doses, inhalation, and man. Too much significance is likely being attached to such experimental data in this age of unwillingness to accept health risks and too little cognizance is taken of the occupational exposure data that have been built up. Plant damage from phytotoxins is much better defined by experience and testing than is animal damage, largely because plants are usually more sensitive to air pollution than are animals and the exposures are more easily controlled.

1. Toxicological Principles

Compounding the difficulty of separating probable cause from multiple exposures are the peculiarities associated with the additivity of effects. If the damage by one pollutant is of the same nature as that from another but the actions of the substances are independent of each other (i.e., the results are simply additive) the pollutants are termed synergistic. If the damage from one pollutant is enhanced by the presence of a second pollutant, the additive effect is potentiated. Potentiation has often been called synergism. If the damage from one pollutant is mitigated by the presence of a second pollutant, the pollutants are antagonistic. Furthermore, increased resistance to a toxic agent can be generated by a prior exposure to that material or to other toxic materials.

There is considerable toxicological evidence to support the idea that there is a threshold concentration below which no physiological damage occurs. As Paracelsus put it in 1530, "Dose alone makes poison." Further backing for the threshold concept is the Arndt-Schulz law which holds that large doses of toxins destroy, moderate doses inhibit, and small doses stimulate. The amount of a toxic material that a physiological system is exposed to will, to a large extent, determine the effects. The amount or dose is usually measured in the product of the airborne concentration (c)

This formulation gives an interesting significance to higher concentrations. It is often assumed that the concentration of material

$$t = k \, e^{-ac} \qquad\qquad\qquad (10\text{-}1)$$

or unconventionally,

$$t \, e^{ac} = k.$$

With a threshold term,

$$t = k \, e^{-a(c - c_t)}.$$

and the time of exposure (t). Haber's law simply states the physiological damage for a given dose is constant, or

$$ct = k. \qquad\qquad\qquad (10\text{-}2)$$

Such an equation cannot hold over a very wide range of variation
for c and t; for example, 3500 ppm of carbon monoxide for 1 hr does
not produce the same effect as 1 ppm for 3500 hr. The equation
can hold in its exactitude only for strictly cumulative toxins.
Haber's law is often modified to accommodate the threshold dose
by the inclusion of another term. For a threshold concentration
it is

$$(c - c_t)t = k, \qquad\qquad\qquad (10\text{-}3)$$

where c_t = threshold concentration.

The use of an exponential relation between dose and damage
gives an equation that can be used over a wider range of variation
for c and t than the hyperbolic relation shown above.
constantly taken into the body (or its metabolite) builds up exponentially with time as

$$C = C_0 \, e^{bt}, \qquad\qquad\qquad (10\text{-}4)$$

where C = concentration of material in the body (or an organ) at
time t (when t = 0, C = C_0) and b = parameter of fit. Equilibrium
is usually assumed when bt equals 5 or 6. The decrease in C
after exposure stops is also assumed exponential and the b
parameter is negative. A term for half-life $(T_{\frac{1}{2}})$ may be defined

as the time for any given C to decrease to one-half its value,
and the equation rewritten as

$$C = C_0 e^{-0.693 t/T_{\frac{1}{2}}} , \qquad\qquad (10\text{-}5)$$

The buildup of a very slowly excreted material such as lead
or arsenic can cause actual poison symptoms of acute disease,
but the usual effect of chronic exposure to very small concentrations is the production of chronic diseases which differ from the
acute diseases produced by short-term exposure to high concentrations.

Principal exposure to airborne materials usually occurs by
respiration--through the stomata of the plant or the lungs of the
animal. Some materials may be deposited on the surface (skin)
and act there or be absorbed through the surface to act as systemic
poison.

Gaseous pollutant materials are inhaled and absorbed into
the bloodstream in much the same manner as oxygen. Transport
into and out of the blood takes place by diffusion across the
pulmonary membrane. The laws of diffusion that govern the
kinetics and equilibrium across membranes can be used to study
the processes involved in exposure to gaseous pollutants.
Soluble gases are much more efficiently removed from the air than
relatively insoluble ones, and very soluble gases may be virtually
completely removed in the nose.

Particles are breathed in with the air and deposited at
varying efficiencies (efficiencies that depend primarily upon
size) in different portions of the respiratory tract. The most
widely used curves are those developed by the International
Committee for Radiation Protection and shown as Fig. 10-1.
These curves are based on all of the available retention data.
The summation curve is a plot of the midpoints of the shaded areas
at each size. The low point in the summation curve results from

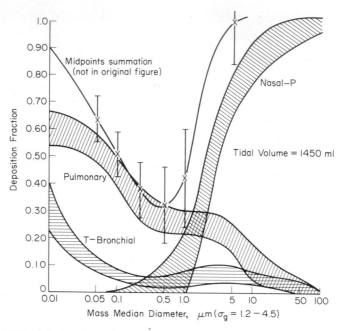

FIG. 10-1. Respiratory retention of particles. From P. E. Morrow et al., Report of the Lung Dynamics Task Group for Committee II of the International Radiological Protection Commission, April 1965.

the fact that particles about 0.2 µm in diameter are not efficiently removed by either the diffusion mechanism (Brownian impaction), which removes small particles, or the impaction mechanism (transport and eddy velocities), which remove large particles. Also, according to Hidy and Brock, diffusiophoresis from water vapor going into the alveolar air retards the diffusive deposition of 0.1 to 0.5-µm particles. In addition to the lung exposure from particulates, slightly soluble particles are usually moved out of the respiratory tract and swallowed. This action gives the effect of ingestion when the actual exposure occurred by inhalation.

EXAMPLE 10-2: What weight of lead particles is inhaled in a day if the concentration is 20 µg/m³? What weight is retained if the effective particle size is 0.2 µ?

Volume of air inspired per day: 20,000 liters (standard man data)

$$\frac{20,000 \text{ liters/day}}{1000 \text{ liters/m}^3} \times 20 \text{ µg/m}^3 = 400 \text{ µg/day}$$

Retained: Efficiency of retention = 35%   (Fig. 10-1)
       400 µg/day X 0.35 = 140 µg/day

2. Correlated Diseases

There have been many epidemiological correlations of disease with air pollution, either directly or indirectly. The indirect correlations have seemingly met with more success than the direct ones. These usually involve correlating the incidence of disease with the size of the city, which in turn correlates very well with the degree of air pollution, both particulate and gaseous. There are, of course, exceptions where different types of fuels are used and where other factors mitigate the general rule that population density accounts for pollution. The direct correlations are generally made by examining the disease rate for certain diseases in areas of high measurements for an air pollution constituent or for several pollutants or general air pollution indicators.

The correlation of smoking, air pollution on the microscale, with respiratory diseases, especially bronchitis and lung cancer, has shown a very strong consistency. Concentrations inhaled by a smoker are orders of magnitude greater than ambient concentrations; for example, the carbon monoxide exposure is about 400 ppm_v, the particulate levels inhaled are in milligrams per cubic meter, and the amounts of other gaseous materials are similarly large.

a. Nonspecific Morbidity and Mortality. The increases in death rates and morbidity during air pollution episodes (catastrophes) have been noted in several major incidents, such as those cited in Chapter 1. Speculations have been made on the specific pollutants responsible, usually with considerable refutation from other investigators who have studied the same data. Glasser and

Greenburg reported at the 1969 American Public Health Association
conference that air pollution, as indicated by sulfur dioxide,
correlates directly with the mortality rate in New York City. They
found that a rise from 0.2 to 0.4 ppm$_v$ of sulfur dioxide was
accompanied by 10 to 20 deaths. Conversely, when the pollution
levels dropped much below normal (sulphur dioxide $\ll 0.2$ ppm$_v$),
the death rate also dropped below normal, which indicates that
normal death rate is related to normal air pollution levels.

   b. Specific Diseases. Of course, the majority of the
correlation efforts have been made to link respiratory diseases to
air pollution. A dramatic 10-year increase to the early 1960s
of approximately 300% in the incidence of emphysema deaths to
about 10 per 100,000 has been attributed to air pollution. A
startling fact about this increase is that it occurred almost
entirely among males rather than throughout the population (more
than fivefold in males). At least part of the increase is attributable to changes in diagnosis; however, total deaths from
respiratory diseases have increased, and it is likely that
changes in diagnosis would not transcend this grouping. Lung
cancer deaths have increased steadily to a level of more than
20 per 100,000. Most of this increase probably resulted from
smoking. Bronchitis is diagnosed much more frequently in
Britain, where it has been correlated with air pollution. Investigation in the United States showed about 20% of the men 40-59
years old and about 15% of all adult males had a productive cough
(bronchitis). Particulate levels have been correlated with
increased incidence of common colds and gastric cancer (see
Section III, A). A prediction that Los Angeles smog would cause
cancer went down in statistical defeat when it was shown that
Los Angeles residents during the 20 years of bad smog had a lower
incidence of lung cancer than the residents of either San Francisco
or San Diego. Heart disease has been linked with the broncho-
constriction related to general air pollution and sulfur dioxide

specifically. These correlations have proved out during catastrophic episodes and later among those who suffered acute
illness during the intense air pollution. Irritability of air pollutants
increases during periods when air pollution determinations are high.

Fallacious correlations must be guarded against. Just the fact
that a correlation exists between an air pollutant and disease does
not mean that the cause-effect relationship has been proved.
Several attorneys are finding this factor in civil suits against
polluters.

c. Symptoms. A survey of physicians, reported in New
Materia Medica in February, 1963, showed 1,600,000 cases
involving air pollution in 1 year. The symptoms, with their
frequencies in percent of patients citing them in parentheses,
were: cough (75.6), smarting of eyes (70.9), tears in eyes (61.5),
nasal discharge (52.0) dyspnea (44.5), sore throat (41.9), chest
constriction (41.7), headache (41.2), choking (33.8), and nausea
(18.2). As may be noted by the sum of the percentages, patients
often complained of more than one symptom.

3. Physiological Responses

For a long time the Russians have used any detectable
physiological response as the basis for setting regulations, while
the United States has relied on physiological damage for the
purpose. The usual result is an order of magnitude difference in
maximum allowables. There is a current trend in this country
toward the Russian philosophy, as evidenced by the carbon
monoxide effects on time interval judgment (see Section III, c).
Such a trend is encouraged by the growing unwillingness of the
public to accept any health risk, an unwillingness fostered by
health officials and the news media.

The physiological change caused by the air pollutant(s)
may be in involuntary functions such as blink rate, respiration,

pulse, and so on, or in performance in psychomotor functions
such as reading rate, tasks that involve judgments of time,
distance, and color, visual acuity, and mental alertness as
determined by other tests. Tests for physical change are often
made on human subjects, especially short-term (a few hours or
less) tests; however, some tests have been made with animals;
for example, the depression of running activity (treadmill revolutions) by dirty ambient air.

4. Ecological Changes

The insults that man makes on his environment are receiving
considerable attention by ecologists. The disappearance of some
species is considered not only a travesty on nature but also an
indication that such could happen to man. Piccard says that it
will happen to man before the end of the twenty-first century.

In addition to the climatic changes previously described,
some stage in the cycle of life may be irreparably damaged. The
most likely portion of the life cycle for such disruption is the
plant. Plants are much more sensitive to air pollution damage
than are animals. The essential role of plants in maintaining
the overall balance of physiological gases in the atmosphere and
their role in the food chain are widely understood. The likelihood
that more than just local problems will occur from an air pollution
caused imbalance in nature seems remote. It is for this reason
that ecologists warn against the graduality of such change as
may lead to the acceptance of the conditions causing the change.
Many ecologists urge responsible people to set meaningful maximum limits and monitor the air to be sure that these limits are not
exceeded. The limits would remain inflexible for all time in this
scheme. An example of the gradual change predictions is one
suggesting that the size of the American desert will increase from
climatic changes brought about by air pollution.

Pronouncements of the doom for mankind come with increasingly frequent regularity. Recently, a newspaper spread the word of yet another expert, a biologist who stated that 35,000 particles/cm^3 will be deadly and reported a level of 15,000 particles/cm^3 that is rising at the rate of 1500 particles/cm^3-year. As a result of this condition, in less than 20 years man will have to wear a gas mask when venturing outside. The date of this article was December 21, 1969, but other days and other agents could be substituted almost ad infinitum. Some speculations refer to the amount of life shortening caused by this agent, or that agent, or pollution in general. Observers who are long-experienced in attempting to measure the effects of ambient air levels on longevity seem more reticent to make definite statements than are newcomers to the field.

D. Economic

The economic effects of air pollution are impossible to assess. Several years ago $60/cap-year was considered a good estimate of air pollution damage in the United States. The number grew to $65 and is still being requoted. Six years ago, the then Surgeon General Luther L. Terry said, "The latest figures suggest that air pollution may be costing the nation $7 billion each year." Ridker estimated $5.5 billion for the cost of air pollution in 1963. His estimate for respiratory disease was $1.989 billion. Linsky wants the figure placed at "over $15 billion/year" and to include a value of all discomforts.

One difficulty that arises in estimating the cost of air pollution is in classifying the extra costs. Are all of the extra costs waste? They create jobs for window washers, building cleaners, car wash establishments, launderers, detergent makers, street cleaners, painters, and so on. Would this money, if it were not spent on air pollution cleanup, be spent in providing useful occupations?

Some observers would like to set a price on the damage
done by air pollution, then give a much smaller figure for the
cost of preventing it. Estimates of both are likely to be grossly
in error. Costs as low as $15/cap-year have been proffered as
the cost for obtaining clean air. The cost of cleaning up auto
exhaust will probably be $250 to $600 per car plus maintenance.
The initial costs of air pollution control devices were $24 on
1969 models, $48 on 1970 models, and $125 on 1971 models.

The uncertainties in assessment of the cost of air pollution
seem to point to control for esthetic reasons. The cost of obtaining
40-mi visibility would be much more than double the cost of obtaining 20-mi visibility because the last fraction of the particles is
much more difficult to remove than the first portion. The cleaning
of gas follows the same pattern of difficulty in removal as particles; that is, low concentrations are hard to remove.

III. EFFECTS OF SPECIFIC AGENTS

The effects of several specific agents or types of agents are
covered in this section. The usual format is to present the effects
in the same order as in Part II; that is, physical, chemical, and
biological. The economic effects are noted only when it is deemed
that worthwhile information is available. When one of the other
three classes of effects is omitted, it may indicate that for that
agent such effects are either absent or insignificant to the total
air pollution damage picture; however, it may only be indicative
of ignorance of the effects.

A. Particulates

The effects of particulates are treated as those effects
contributed to airborne particles in general, without generic
identification, but some specific types of particles are covered.
The physical effects are those with the soundest basis; some of
the chemical and biological effects attributed to particulates are
debatable.

1. Physical Effects

The important physical effects of particulates are reduction in visibility and isolation, increase in rainfall, and dirtiness, which was sufficiently described in Part II, except that it might be pointed out that archeologists usually dig to find the remains of old civilizations.

a. Visibility Reduction. Visibility is limited by light scatter and absorption by gaseous molecules (Rayleigh scatter) and suspended particles (Mie scatter) according to the principles cited in Cahpter 5. Visibility is the mean greatest visual range that persists over 50% of the horizon, visual range being the maximum distance at which a large black object against a light background can be seen sufficiently for identification.

The contrast between two lines of sight (object and background) is defined by

$$C = (I - I_0)/I_0,$$ (10-6)

where C = contrast = -1 for black object ($I = 0$), I = illumination of object, and I_0 = illumination of background. Light scattered into the line of sight to the dark object reduces the contrast. Because the probability of interaction between the photon and the constituents of its medium is random with respect to distance of travel, the contrast is attenuated exponentially with distance; that is,

$$C = C_0 e^{-\beta x},$$ (10-7)

where C_0 = contrast at the object ($x = 0$), C = contrast at distance x from object, and β = attenuation coefficient. The attenuation coefficient (β) is often broken into parts: $\beta = b + k$, where b is the scattering coefficient and k is the absorption coefficient. Both of the factors can vary over wide ranges. The values of b range from several ft^{-1} for a dense fog to 30 ft^{-1} for a clear day, and the values for k may be less than 1 ft^{-1} for dense, black

smoke or nearly infinite for nonabsorbing gaseous materials,
such as nitrogen and water vapor, or particles, such as water.
The attenuation or illumination may also be expressed as

$$I = I_0 e^{-\beta x} = I_0 e^{-(b + k)x}.$$ (10-8)

The contrast limen (e) is the contrast at the visual range (v).
This definition applied to Eq. 10-2 gives

$$\epsilon = -e^{-\beta v}.$$ (10-9)

ϵ is often used at -0.02 for standard visibility (theoretical) or
as -0.05 for meteorological visibility (practical). The latter
value gives

$$v = -\frac{1}{\beta} \ln (-\epsilon) = \frac{2.996}{\beta}.$$ (10-10)

If an effective scattering cross section (K_s) is defined as
the ratio of the effective cross-sectional area of particles,

$$\beta = K_s N \pi d^2 / 4,$$ (10-11)

where K = effective scattering cross section, N = number of
particles per unit volume, and d = diameter of particles. For
various diameters of particles, if the diameters are divided into
n groups, each characterized by a diameter d_i,

$$\beta = \frac{\pi}{4} \Sigma_i K_{si} N_i d_i, \qquad i = 1 \text{ to } n.$$ (10-12)

Since $N = W/\rho V = W/(\rho \pi d^3/6)$, where W = weight of particles
per unit volume, ρ = density of particles, V = volume of a
particle,

$$\beta = 3K_s W/(2\rho d).$$ (10-13)

W and ρ should be consistent in units for d to be the reciprocal
unit of β. Particle diameter is usually put in dimensionless form
by defining a term α $(\alpha = \pi d/\lambda)$, where λ = the wavelength of the
interacting photon.

$$\beta = 3\pi K_s W/(2\rho\lambda d)$$ (10-14)

The λ value used for characteristic sunlight is 524 nm. Values of K_s for different indexes of refraction (m) are shown plotted against α in Fig. 10-2. When K_s is not known, it is common to use a value of 2.

For plume opacity calculations, Eqs. 10-8 and 10-14 are combined as

$$\ln I/I_0 = -\beta x = \frac{-3x}{2\rho} \Sigma i \, K_{si} W_i/d_i \, , \, i = 1 \text{ to } n \qquad (10\text{-}15)$$

Eq. 10-15 is for n size groups.

Various investigators have proposed a rule of thumb relation in which the weight of particles in a square meter column of a length sufficient to extinguish visibility is a constant (for relative humidity less than 70%).

$$R_v = k/W, \qquad (10\text{-}16)$$

where R_v = visual range (m), k = weight in a square meter column for length R_v (gm/m^2), and W = mass concentration in air (gm/m^3). This is by no means an exact relation. Early estimates of k for fog and smoke were about $0.5 \, gm/m^2$, but current articles report a value of $1.8 \, gm/m^2$. Charlson reports a value of $0.45 \stackrel{x}{\div} 2 \, gm/m^2$ for the extinction coefficient (W/b) alone; therefore the $1.8 \, gm/m^2$ for k in Eq. 10-16 means that R_v would be for x = 4, since b + k, or exp −4 is taken as 0. In addition, the typical ratio of b for particles to k for nitrogen dioxide is 7 at 500 nm and 10 at 550 nm, and k is $0.7 \, ppm_v^{-1}/ km^{-1}$.

EXAMPLE 10-3: What will be the visibility in an atmosphere containing $0.3 \, ppm_v$ of sulfur dioxide in equilibrium with sulfuric acid particles formed from the sulfur dioxide if it is assumed that the lifetimes of the sulfur dioxide and the sulfuric acid particles are the same?

$$0.3 \, ppm_v \text{ of sulfur dioxide} = 0.3 \, cm^3/m^3 \times 64 \, gm/gm \text{ mol}$$
$$\times 1/24,100 \, cm^3/gm \text{ mol}$$
$$= 8 \times 10^{-4} \, gm/m^3$$

Weight of sulfuric acid: $98/64 \times 8 \times 10^{-4} = 12 \times 10^{-4} \, gm/m^3$

$$R_v = k/C = (1.8 \, gm/m^2)/12 \times 10^{-4} \, gm/m^3 = 1500 \, m$$

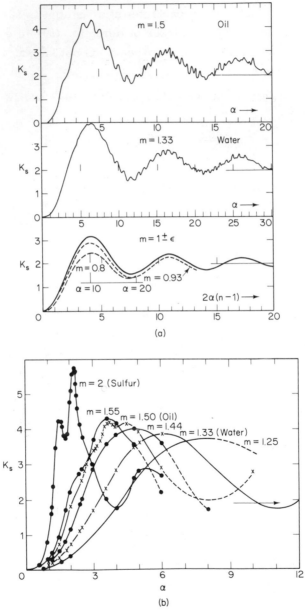

FIG. 10-2. Effective scattering cross sections. (a) Three cycles. (b) Smoothed first cycles. From H. C. van de Hulst, <u>Light Scattering by Small Particles</u>, Wiley, New York, 1957.

EXAMPLE 10-3: Calculate the absorbance (optical density) and the equivalent Ringelmann number for a plume which carries 3220 lb/day of aluminum silicate catalyst in 174,000 cfm of flue gas from a stack 2.66 m in diameter. The particle density is 2.466 gm/cm3 and the size distribution is:

Size (μm) Range	Midpoint	Percent in range	Concentration c, (gm/m³)	$2\alpha(m-1)$ $= 7.68\,d$	K_s	$\dfrac{K_s - c}{d}$
>45	50	1.50	0.0031	384	2[a]	0.0001
20-45	32.5	1.30	0.0026	250	2[a]	0.0002
15-20	17.5	4.40	0.0091	134	2[a]	0.0010
10-15	12.5	17.80	0.0367	96	2[a]	0.0059
5-10	7.5	19.50	0.0402	58	2[a]	0.0107
4-5	4.5	15.50	0.0319	34	2[a]	0.0142
3-4	3.5	11.50	0.0237	27	2[a]	0.0135
2-3	2.5	13.50	0.0278	19	2	0.0222
1-2	1.5	11.50	0.0237	12	2.2	0.0347
<1	0.5	3.50	0.0072	3.8	3.2	0.0462

$$\Sigma = 0.1487$$

[a] K_s is assumed equal to 2 when $2\alpha(m-1) > 20$.

Concentration: Total $\dfrac{3220 \times 454}{(174{,}000/35.3)1440} = 0.206\ \text{gm/m}^3$

m = 1.64 (for aluminum silicate, andalusite, Al_2SiO_5, from Handbook of Physics and Chemistry)

$\ln I/I_0 = -3(2.66)/[2(2.466)]\,(0.1487) = -0.241$ $I/I_0 = e^{-0.241} = 0.80$

Absorbance $= \log 100/\%T = \log 100/80 = 0.0997$ Equivalent Ringelmann $= (100 - 80)/20 = 1.0$

Theoretical calculations of the particle sizes most effective
at visibility attenuation yield sizes near the wavelength of the
light. The wavelength of daylight used often is 524 nm. Figure
10-3 shows a plot of the weights for various sizes and types of
particles at the extinction limit.

The effective diameter of most ambient suspended particle
populations (after aging) is about 0.6-0.8 μm. Charlson reports
a size of 0.682 μm effective diameter for Los Angeles smog
particles. It is noted that the size of ambient particulates
approaches the size of maximum attenuation efficiency.

Burt observed an exponential relationship between the
visibility and soiling index in St. Louis. The relation was
approximately

$$R_V = 11e^{-0.479 \text{ S.I.}}, \qquad (10\text{-}17)$$

where R_V = visual range (mi) and S.I. = soiling index (Cohs).

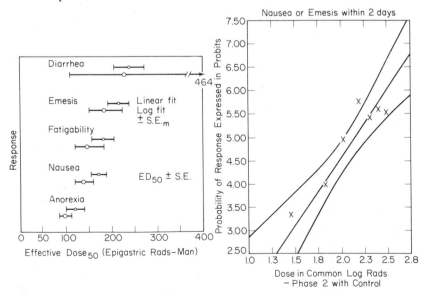

FIG. 10-3. Response to radiation. From <u>Fundamental
Nuclear Energy Research--1966</u>, U.S. Atomic Energy Commission,
Washington, D. C., December 1966.

b. Other Physical Effects. Reduced insolation can be
calculated in the same manner that visibility is, except that
the scatter of a photon does not necessarily keep it from reaching the earth. Probably about one-half of the b value adds to
the total k value to make up the β value for insolation reduction.
The distance of attenuation is usually rather small, less than
$\frac{1}{4}$ mi. The layer involved is less than or equal to the mixing
depth. Insolation reduction by particulates is often 10-15% and
reaches 35-50% in some cases. Before Pittsburgh was cleaned
up, particulates caused "night" at midday. Volcanic particles
have normally been cited as causing the temperature drop noted
by some observers.

Although increased rainfall in some polluted atmospheres
over cities is sometimes attributed to particulates, the more
logical explanation is that increased convection caused by the
"heat island" is responsible for the increased rainfall. Some
observers go so far as to say that air pollution particles decrease
rainfall because they provide too many condensation nuclei.

2. Chemical Effects

Particles in ambient air are quite small and have a large
surface area-to-mass ratio. For example, 0.7-μm-diameter
spherical particles with a density of 2.65 gm/cm^3 have a surface
area of 3.43 X 10^4 cm^2/gm of material. This property results in
many of the chemical effects attributable to particles. Small
particles are more soluble than large particles, more combustible,
and more mobile in renewing surface reactants by means other than
diffusion.

Flash fires and explosions are caused by dense suspensions
of combustible dusts. Such problems are limited mostly to enclosed spaces, the most notorious being grain elevators and coal
mines. However, rather rapid oxidations of small particles occur
in the atmosphere without the particle density being sufficient
to support flame propagation.

Particles usually react with surfaces only in the presence
of moisture; however, significant capillary condensation occurs
under particles when the relative humidity is far below saturation,
and some of the most damaging particles are hygroscopic and
take up moisture from the surroundings and dissolve at atmospheric humidities. Air Quality Criteria for Particulate Matter
notes adverse effects on steel and zinc panels when the suspended particulate concentration exceeds 60 $\mu g/m^3$ for a yearly
average in the presence of sulfur dioxide and moisture.

3. Biological Effects

Particles may be classified into three categories according
to their biological action: inert particles such as limestone,
carbon, and talc; irritating shapes such as fibers of glass, rock
wool, and bagasse; and chemically active particles such as
silica, beryllium, and asbestos.

Dusts deposit on plants and may interfere with the reception
of sunlight or with plant functions such as respiration and transpi-
ration. Inert dusts result in the stunted growth of plants that
receive large amounts of such dust. Soluble particles and mists
often react with plants and cause biochemical damage and even
plant death; for example, sulfuric acid mist is an effective defoliant.

The biological effects of inert dusts on people and other
animals may be classed as mechanical blockage. The lung-
cleansing mechanism can be overwhelmed or made ineffectual,
causing a buildup of dusts deposited in the lungs during respiration. Under normal circumstances inert dusts can be respired
with apparent impunity.

Irritating dusts may cause problems because of their
shape, chemical composition, radioactivity, or combination of
these factors. The most irritating shapes are fibers, such as
rock wool and fiber glass. Particles that cause the most concern

chemically are probably silica and beryllium. Asbestos is an irritating shape and material. Radioactive particles may deliver large dosages of radiation to small areas. The highly active particles that would result if a nuclear spacecraft were destroyed in the atmosphere have been the subject of considerable research.

a. Asbestos. The linkage of asbestos with some lung cancers (mesotheliomas) has resulted in a rapid increase in the number of publications on the subject. Apparently, certain polycyclic hydrocarbons must be present for the formation of such cancers and the asbestos may just potentiate the effects of the hydrocarbons. Before the indictment of asbestos as carcinogenic, the concern was for asbestosis, a pneumoconiosis with fibrotic manifestation.

b. Beryllium. Beryllium can cause berylliosis--a morbid condition of the lungs characterized by the formation of granulomas. A report of numerous cases of berylliosis around a plant that gave off low levels of beryllium gave rise to the adoption of very low levels for the TLV (0.02 $\mu g/m^3$), and the American Industrial Hygiene Association recommends a maximum ambient level of 0.01 $\mu g/m^3$ over 1 month. It is noted that the maximum ambient concentration recommendation is much closer to the TLV than is usually customary--1/2 versus the usual 1/10 or even 1/30. There is considerable belief that the basis for TLV is erroneous and that 0.05 $\mu g/m^3$, or perhaps 0.15 $\mu g/m^3$, would be a safe occupational level.

4. Inert Dusts

The Air Quality Criteria of the National Air Pollution Control Administration (NAPCA) recommend maximum ambient levels of 80 $\mu g/m^3$ suspended particulates as an annual mean for health reasons. The AIHA recommendations for an air basin average over a period of 30 days are: suspended particulates, 75 $\mu g/m^3$; dustfall, 1.0 mg/cm^2; soiling index, 0.4 Coh. For single-point measurements over 30 days the recommendation is: suspended particulates, 120 $\mu g/m^3$; dustfall, 1.5 mg/cm^2; soiling index, 0.5 Coh.

Iron oxide is an inert dust with a TLV of 10 mg/m^3 to prevent changes in the lung x ray, changes that are not necessarily harmful. The AIHA ambient air recommended maximum limits , for a 24-hr average based strictly on nuisance, are 100 $\mu g/m^3$ in rural areas to 250 $\mu g/m^3$ in industrial areas, or 100 $\mu g/m^3$ in an industrial area for a 30-day average.

A study in the Buffalo area resulted in a correlation of gastric cancer incidence with suspended particulates, the high-concentration areas having about twice as many gastrointestinal cancers as the cleaner areas.

B. Sulfur Dioxide

Although there is a large amount of literature on the effects of sulfur dioxide, much of it is contradictory. Some of the effects are evidently rather well agreed upon, for example, the important role of sulfur dioxide-derived particulates in visibility attenuation and the potency of sulfur dioxide as a phytotoxicant. However, the animal effects should best be dealt with under the epidemiological umbrella of general air pollution rather than the more specific toxicological tests when it is desired to show probable physiological damage.

Sulfuric acid and other sulfates have been reported at usual levels of 5-20% of the total urban suspended particulates. The particles are rather small ($\geq 80\%_w$ are < 1 μm) and are, therefore, highly efficient at visibility attenuation. Various rates of sulfur dioxide conversion have been reported, such as 0.1%/hr (lifetime of 6 wk) and 1%/hr (4 days).

Damage from sulfur dioxide occurs among sensitive plants at long-term concentrations of 0.1 ppm_v or at a 1-hr concentration of 1 ppm_v. Incipient damage to alfalfa, for maximum damage values of concentration (c) and time (t), may be represented by $c_t = 0.33$ k = 0.92 in Eq. 10-2 and 100% leaf destruction with $c_t = 2.6$ and k = 3.2. Thomas and Hendricks show linear equations of the yield as percent leaf destruction for alfalfa, wheat and barley. Without control or tall-stack dilution around smelters, sulfur dioxide has often completely denuded large areas of vegetation and affected sensitive plants to distances of several miles.

Amdur's work with guinea pigs has shown that sulfur dioxide causes increased airway resistance in guinea pigs--from 10% increase at 0.16 ppm_V to 265% increase at 835 ppm_V. The effect was potentiated 3 to 4 times when the sulfur dioxide was accompanied by soluble catalysts of ferrous iron, manganese, and vanadium at concentrations of 700-1000 $\mu g/m^3$. This changed the mildly irritant sulfur dioxide gas to strongly irritating droplets of sulfuric acid. Such research does not suggest that serious effects on animals result from ambient sulfur dioxide exposures.

Guinea pigs exposed for 1 year to 0.1 ppm_V and 1 ppm_V at Hazleton Laboratories showed no physiological damage. Those exposed to 5 ppm_V for the same period had a significantly higher survival rate and showed fewer lung lesions than the controls, indicating perhaps that the Arndt-Schulz law is operative. The Air Quality Criteria for Sulfur Oxides concludes that increased mortality from bronchitis and lung cancer may be caused by a sulfur dioxide concentration of 0.04 ppm_V annual mean with smoke of 160 $\mu g/m^3$, and increased mortality (20%) may occur from 0.52 ppm_V sulfur dioxide (24-hr concentration) with suspended particulates to give a soiling index of 6 Cohs (about 2000 $\mu g/m^3$). These conclusions and that of Glasser and Greenburg (p. 375) are based on the presence of increased sulfur dioxide concentrations during the period of air pollution episodes that showed increased mortality. Many other factors are also changed under these air pollution conditions and some of them, whether measured or not, could be responsible for the increased deaths noted.

C. Carbon Monoxide

The important effects of carbon monoxide are the biological effects on people and animals. There have been a great number of studies made on carbon monoxide effects, levels, chemistry, and so on. The bibliographic references run into the thousands. Using the estimated United States releases as a basis, one might rate worldwide emissions at more than 200 million tons/year, or uniformly dispersed, about 0.03 ppm_v. Studies conducted by various observers and means indicate a half-life of approximately 1-6 months for carbon monoxide in the atmosphere.

Many of the physiological studies on carbon monoxide have involved the uptake and release kinetics and the amounts of hemoglobin inactivated by various concentrations. Carbon monoxide reacts with hemoglobin (Hb) to give carboxyhemoglobin (COHb). Not only is the Hb unavailable for oxygen transport but also the dissociation of oxygen from the other Hb molecules is shifted to a slower rate.

The Haldane equation is used to express the steady-state equilibrium condition for carbon monoxide.

$$[COHb]/[O_2Hb] = R \; PP_{CO}/PP_{O_2}, \qquad (10\text{-}18)$$

where PP = partial pressure, R = equilibrium ratio or relative affinities of carbon monoxide and oxygen for Hb, and brackets indicate concentration. The equation is normally used to calculate the percent of COHb, or

$$Percent \; COHb = 100 \; R \times PP_{CO}/(PP_{O_2} + R \times PP_{CO}) \quad (10\text{-}19)$$
$$= 100 \; PP_{CO}/(PP_{O_2}/R + PP_{CO}).$$

R is often quoted as 210; however, the most reliable data apparently show that a carbon monoxide concentration of 30 ppm_v

gives about 5% carboxyhemoglobin in 8-12 hr, and 200 ppm_v
causes about 20% in 6-8 hr. The corresponding R values are
368 and 262, respectively. Perhaps R should be taken as 300.

Sometimes percent carboxyhemoglobin is assumed equal
to 1/6 times the carbon monoxide concentration in parts per
million by volume. At any rate, R apparently varies with the
pH of the blood, the temperature, and the individual.

The significant exposures of people to carbon monoxide are
caused by auto exhaust (approximately 30 ppm_v for 8 hr), cigarette
smoking (400 ppm_v inhaled during smoking), and physiological
function, hemoglobin catabolism (~0.4% carboxyhemoglobin).
Physiological symptoms have not been proved for less than 10%
carboxyhemoglobin; therefore current efforts mostly indicate
physiological change for low levels. It has been reported that
about 2% carboxyhemoglobin affects the ability to judge time
and space intervals and 5% impairs ability to perform designated
psychomotor activities. Heavy smokers often have levels above
5%.

The effects of carbon monoxide are additive with other
conditions that cause hypoxia; for example, 7000-ft elevation
may be equivalent to about 5% carboxyhemoglobin.

D. Lead

Lead exposure from the atmosphere is less than one-sixth
as much as that from diet, food, and beverage; the average
intake values are about 15 mg/year from air, 5 mg/year from
water, and 100 mg/year from food. However, the efficiency
of absorption may be 5 times as high for the air intake as for
that in the diet (about $50\%_w$ versus $10\%_w$), making the air
exposures approximately $40\%_w$ of the total absorbed. Most of
the atmospheric lead comes from leaded gasoline, about
500 million lb/year. Urban concentrations average only about
3-4 $\mu g/m^3$, but they may run as high as 50 $\mu g/m^3$ near express-

ways. A considerable portion of the diet lead probably originates
with lead from gasoline which is taken up by plants and dissolved in liquids.

Much of the recent concern about atmospheric lead was
started by Patterson. The rebuttal came from data of long-term
industrial exposures at 150 $\mu g/m^3$ without apparent ill effect
(TLV = 200 $\mu g/m^3$). The average blood lead levels vary by a
factor of <3 worldwide and some of the high values are among
people living far from large numbers of automobiles. Despite
the lack of toxicological indictment of lead and the efforts of
the lead industry, opposition to leaded gas will probably continue until a changeover to unleaded gas takes place. Only a
very small portion of the gasoline produced now is unleaded.

The AIHA has recommended maximum allowable ambient
lead concentrations of 10 $\mu g/m^3$ for a 30-day average. Arizona,
part of California, and others have set standards of 10 $\mu g/m^3$;
however, Montana, Pennsylvania, New York City, and Dallas
set their maximum levels at 5 $\mu g/m^3$.

E. Photochemical Smog--Ozone, Nitrogen Dioxide, PAN, and Polycyclics

Smog and its specific components are quite efficient in
producing physical, chemical, and biological effects. The
visibility reductions of the Los Angeles smogs have been pictured
repeatedly in papers, magazines, and journals; the ability of
smog to cause chemical damage to rubber, clothing (fading of
color and disintegration or weakening of fiber, especially nylon
stockings), paints, and exposed surfaces has often been described; and the high efficiencies of smog in causing biological
damage to plants and animals have been documented in many
tests.

Ozone is both biologically active and present in significent concentrations. It is present at concentrations that average about $0.15-0.20$ ppm_V during 24 hr of heavy smog and may peak around 1 ppm_V. Ozone accounts for about 90% of the oxidant measured by the potassium iodide test. Ozone causes damage to tobacco leaves at concentrations as low as 0.02 ppm_V, which is about the odor threshold for most people. Many plants are severely damaged by exposures below 1 ppm_V for a few hours, and 3 ppm_V causes damage in a short time. Pinto beans showed leaf damage from 0.2 ppm_V for 2 hr. Toxicity experiments on animals have produced bronchitic pneumonia, edema, and mortality at concentrations of $0.1-3$ ppm_V. An increase from 38 to 85% in the number of lung tumors in a tumor-prone strain of mice was caused by 15 months of 1-ppm_V exposures. These ozone effects data are mitigated somewhat in their import because of the antagonism or cross-resistance between ozone and many other toxic pollutants. For example, the presence of sulfur compounds such as hydrogen sulfide and sulfur dioxide, which are themselves toxic, has protected mice at ozone concentrations that would otherwise have been lethal. Ozone causes many of the same reactions as ionizing radiation-- oxidation, cross-linkage of molecules, mutations, and apparent premature aging. The ozone layer is of concern for high-flying aircraft. Recommended ambient standards for ozone are normally based on plant damage considerations and may be around 0.05 ppm_V for a few hours.

Nitrogen dioxide causes visibility attenuation which may become severe at concentrations of several tenths of 1 ppm_V. It causes overt plant damage at a few parts per million by volume, but may show growth reduction for some plants and fruits at a level of 0.5 ppm_V or less. High concentrations, a few hundred parts per million by volume, are highly toxic to man. Nitrogen

dioxide has an odor threshold around 2 ppm_v and a distinctive
orange color. The effects of low concentrations are usually not
separable from those of ozone since they occur together. These
effects include increased airway resistance, increased susceptibility to pneumonia, and hardening of the lung tissue. The
California standard of 0.25 ppm_v for 1 hr is based on visibility
reduction.

PAN is usually present in photochemical smogs at levels
of about 50 ppb_v and has been measured at 210 ppb_v in Los Angeles.
A concentration of 14 ppb_v for 4 hr has caused visible plant damage.
PAN is very reactive and is one of several such types in smog;
there are other acyl groups in addition to acetyl, such as propional,
butyryl, and benzoyl. Ozonated olefins and free radicals, especially of peroxides, are present in smog. PAN is highly efficient
as a lachrymator (an estimated 200 X that of formaldehyde), but
the eye irritation of the smog is caused by other substances as
well. Peroxybenzoyl nitrate (PBzN) may be as important in its
physiological damage as is PAN.

There are many polycyclic hydrocarbons among ambient
pollutants. The ones that have received the most attention are
benzo[a]pyrene and benzo[e]pyrene. Polycyclics from roofing
pitch have been identified as the causative agents in cancer of
the scrotum and other skin cancers among roofing workers. This
has been the basis of concern regarding the polycyclics.

F. Fluorides

Most inorganic fluorides show chronic toxicity to plants
and animals. Toxic levels of fluorides are quite low and are
reached by emissions from several types of industries, including
phosphate fertilizer and detergent builder operations (emissions
during the mining of and dissolution of phosphate rock), aluminum
processing with cryolite, brick plants, pottery and ferroenamel
plants, and other metals industries processes. The gaseous

fluorides (HF, SiF_4, H_2SiF_6, and so on) are the most efficient in causing plant damage and etching of glass and the particulates [$CaFCa_4(PO_4)_3$--apatite, Na_3AlF_6--cryolite, CaF_2--fluorspar, FeF_3, NaF, and so on] are the most damaging to foraging animals.

Fluoride phytotoxication is evidenced by necrosis of the edges of the leaves. Some of the most sensitive plants are gladioli and certain fruit trees, such as prune, apricot, and peach. These plants are damaged by as little as 0.02-0.05 ppm_v.

Grazing animals show chronic damage when their total diet contains more than about 40 ppm on a dry weight basis. Sodium fluoride is about twice as toxic as the relatively insoluble forms of apatite, cryolite, and fluorspar. The saliva of the animal converts gaseous fluorides to sodium fluoride. The symptoms are those of starvation which are caused by fluorocitrate blockage of the Kreb's cycle plus fluorosis which results in a wearing away of the teeth. The fluoride particles may settle out on the leaves and cause animal toxicity without damaging the plants. Man can show similar symptoms, but poisoning of animals is caused much more frequently.

The presence of gaseous fluorides etches glass in neighborhoods where such fluorides are released. The presence of a water film accelerates the etching.

The AIHA recommends maximum levels on forage (as fluorine on a dry weight basis of: 40 ppm for a yearly average; 60 ppm for 2 consecutive months; 80 ppm for any month). They recommend that gaseous levels be kept below the following values: 4.5 ppb_V average for 12 hr; 3.5 ppb_V average for 24 hr; 2.0 ppb_V average for 1 week; 1 ppb_V average for 1 month.

G. Arsenic

The classic problem of atmospheric arsenic poisoning was the case in which sheep and horses died from ingesting grass

contaminated by the dust from a copper smelter; both the fresh
grass and hay caused animal deaths. Most of the atmospheric
concentrations today are from ore smelting and pesticides used
in agriculture. One important source which is on its way out is
the burning of hulls and motes at cotton gins. Cotton often has
arsenic residues from arsenic acid which is used as a defoliant
preparatory to mechanical harvesting.

European exposures, mostly lead and calcium arsenates,
have resulted in reported carcinogenesis; however, the United
States experience, mostly arsenic trioxide, has not caused
cancer formation. Furthermore, arsenic has been added to poultry
and cattle feeds to counteract selenium toxicity. A story of
arsenic addiction by people near a smelter has been reported,
complete to the withdrawal symptoms.

The TLV for arsenic is 0.5 mg/m^3. The ambient concentrations reported by the NASN for 1964-1965 at 133 urban stations
averaged only 0.02 μg/m^3.

H. Miscellaneous Agents

The effects of a few of the many airborne pollutants not
previously covered are described briefly here; these are ethylene,
aldehydes, pesticides, allergens, and odors.

1. Ethylene

Ethylene affects the metabolism of a number of plants,
causing damage to leaves, flowers, and fruit produced by the
plants. Ethylene has been used in concentrations of 75-90%$_v$ in
oxygen as an anesthetic. It is apparently not toxic to animals.
Atmospheric concentrations as low as 0.04 ppm$_v$ have caused
plant damage. Cotton, orchids, carnations, snapdragons, and
roses are all sensitive to ethylene. The AIHA-recommended
maximum allowable ambient concentrations are based on zones,
and the values for rural, residential, commercial, and industrial,

respectively, are: 0.25, 0.50, 0.75, and 1.00 ppm_v for 1-hr averages, and 0.05, 0.10, 0.15, and 0.20 ppm_v for 8-hr averages. The California standards are 0.5 ppm_v for 1 hr and 0.10 ppm_v for 8 hr.

2. Aldehydes

Aldehydes have often been blamed for the odors of diesel exhaust and for the eye irritation of photochemical smog. Vogh has shown that lower aldehydes are not responsible for diesel odors, and perhaps none are. The eye irritation caused by smog is now attributed to many materials, several of which are much more effective than formaldehyde. Aldehydes are toxic at higher concentrations, but they are not believed to be toxicologically significant at present ambient levels (usually less than a few tenths of 1 ppm_v but occasionally as high as 2 or 3 ppm_v, as formaldehyde). The AIHA-recommended maximum allowable ambient concentrations are 0.1 ppm_v of formaldehyde, 0.01 ppm_v of acrolein, and 0.2 ppm_v total aldehydes.

3. Pesticides

Pesticides released into the atmosphere are rapidly diluted below toxic levels except as allergens. Herbicides and defoliants have caused problems when they have drifted onto adjacent fields. The principal concern about pesticides is that they may be reconcentrated in water or ground and cause ecological upsets by eliminating sensitive species of birds, bees, and other animals. Some of the airborne levels of pesticides, toxaphene in particular, in cotton fields across the south are sufficient to cause distress to sensitive people.

4. Pollen and Other Allergens

Allergens cause a sensitized reaction in hay fever sufferers, asthmatics, and others (about 10% of the population). The reaction is a release of histamines, swelling of the mucous mem-

branes of the respiratory system, and the attendant sneezing
and breathing difficulties. Pollen is indeed a serious air pollution problem. The presence of several particles per cubic meter
is enough to bother many people. The billions of particles
(mostly 10-50 μm) per pound and the 1.6 million lb/year of
pollen released in the United States assure allergenic problems.
Ragweed is the most notorious source of hay fever allergen, but
particles from grasses, oaks, cottonwoods, alfalfa, cedars,
elms, and many other organic sources cause trouble. Some of
the major incidents have occurred in Yokohama, New Orleans,
and Minnesota.

5. Odors

Odors cause psychological changes which can affect the
physical well-being. This premise has been upheld in the courts.
The usual direct problems are anorexia, nausea, and hypertension. Even pleasant odors become obnoxious when they persist
at a high intensity for a long period. The odors that cause most
complaints are acrolein (cooking fat), skatole (manure), putre-
scine (decaying putrescible matter), cadaverine (dead animals),
and the mercaptans (skunk musk, butyl; paper mill, methyl and
ethyl). Disagreeable odors are generally amines, sulfur compounds, and organic acids (valeric, body odor; butyric, dirty
feet); pleasant odors include esters (fruity odors), alcohols,
and perfumes.

I. Noise

Noise is unwanted sound. It is considered by many as
our number one pollutant. For a long time only the hearing loss
caused by noise (presbycusis) was described in a definitive
manner. All other effects were vaguely alluded to as psychological effects. Noise causes hypertension, dilation of the pupils,
drying of the mouth, contraction of the muscles and blood ves-
sels, excess production of adrenaline, stoppage of gastric
juice flow, and excitation of the heart. Meaningful correlations

exist between heart disorders and noise. The discomfort threshold for noise is about 117 dB at 2000-5000 Hz, and the pain
threshold is about 143 dB (both based on sound pressure level,
L_p). Attempts to relate physiological response have resulted in
the defining of sones and phons, but these units have not been
popularly used. A sone is the loudness of 1000 Hz tone 40 dB
above listener's threshold. The loudness level in sones is the
number of multiples of the 1-sone intensity in the judgment of
the listener. The loudness in phons is the 1000 Hz tone in dB
that is judged to have the same loudness as that being rated.

Eighteen million Americans suffer from some hearing loss.
They will be joined within 10 years by 50% of those being exposed to 95 dB of noise for 40 hr/week. These are permanent
hearing losses; there are many more temporary losses of hearing.
The AIHA recommendations for maximum noise exposures generally
parallel those shown in Table 7-4. These levels are based on
hearing loss risk (see Table 10-2) and other factors.

I. Radiations

The effects of ambient levels of ionizing radiations, photons, and corpuscular radiations, have been dramatically overemphasized. It is true that ionizing radiations can cause cancer,
mutations, and life shortening. It is doubtful that the amount of
fallout from nuclear releases by atomic testing of weapons and
by reactors (a total in the United States of about 1 roentgen
exposure; 1 roentgen releases about 86 ergs/gm in air) has caused
any significant physiological damage. There were some higher
exposures near the test sites that caused real effects, especially
beta burns to natives of Rongelap and to horses and cattle in
Nevada.

Experimental data indicate that significant numbers of the
effects that have received the most attention (cancer, mutations,
and life shortening) are caused by doses many times those levels

TABLE 10-2

Percentage Risk of Developing a Hearing Handicap[a]

Percentages of exposed population

Exposure level (dBA)	Age Exposure (age − 20)	20	25	30	35	40	45	50	55	60	65 Years
	Exposure	0	5	10	15	20	25	30	35	40	45
80	Total	0.7	1.0	1.3	2.0	3.1	4.9	7.7	13.5	24.0	40.0
	Noise-induced	0.0	No increase in risk at this level of exposure								
85	Total	0.7	2.0	3.9	6.0	8.1	11.0	14.2	21.5	32.0	46.5
	Noise-induced	0.0	1.0	2.6	4.0	5.0	6.1	6.5	8.0	8.0	6.5
90	Total	0.7	4.0	7.9	12.0	15.0	18.3	23.3	31.0	42.0	54.5
	Noise-induced	0.0	3.0	6.6	10.0	11.9	13.4	15.6	17.5	18.0	14.5
95	Total	0.7	6.7	13.6	20.2	24.5	29.0	34.4	41.8	52.0	64.0
	Noise-induced	0.0	5.7	12.3	18.2	21.4	24.1	26.7	28.3	28.0	24.0
100	Total	0.7	10.0	22.0	32.0	39.0	43.0	48.5	55.0	64.0	75.0
	Noise-induced	0.0	9.0	20.7	30.0	35.9	38.1	40.8	41.5	40.0	35.0
105	Total	0.7	14.2	33.0	46.0	53.0	59.0	65.5	71.0	78.0	84.5
	Noise-induced	0.0	13.2	31.7	44.0	49.9	54.1	57.8	57.5	54.0	44.5
110	Total	0.7	20.0	47.5	63.0	71.5	78.0	81.5	85.0	88.0	91.5
	Noise-induced	0.0	19.0	46.2	61.0	68.4	73.1	73.8	71.5	64.0	51.5
115	Total	0.7	27.0	62.5	81.0	87.0	91.0	92.0	93.0	94.0	95.0
	Noise-induced	0.0	26.0	61.2	79.0	83.9	86.1	84.3	89.5	70.0	55.0

[a]From A. Glorig, "Hearing Conservation—The Need for It," Noise News Rev., 1:1, 9–10, February 1970.

that have existed; for example, the dose of radiation required to
double the natural mutation rate is 30-200 roentgens. Small
doses of radiation, just as small doses of chemical toxins, have
been shown to lengthen life of test animals and to stimulate
growth of plants and animals. In addition, the premise that all
radiation damage accumulates has been refuted by evidence that
acute radiation damage is repaired at an exponential rate with a
half-repair time of several days. The median lethal dose (LD_{50})
may be about 600 roentgen equivalents of penetrating radiations.
Sublethal doses cause symptoms of bone marrow and gastrointestinal damage (see Fig. 10-3).

The maximum permissible occupational dosage of ionizing
radiation recommended by both the NCRP and the ICRP is about
5 rems/year or 100 mrem/week. [The absorbed dose in rems is
calculated by multiplying the absorbed dose in rads (1 rad =
an absorbed dose of 100 ergs/gm) by the relative biological
effectiveness (relative to 200 kV x rays) of the radiation in
question.] The allowable public exposures are generally 1/10
to 1/30 of the occupational exposures dependent upon the number
and ages of the people exposed.

The biological dangers from other radiations include sun-
burn (UV), eye damage, and mutations. Radiations may come
from the sun, lasers, welders, radar, and electronic ovens.

IV. SUMMARY

There are many effects attributable to air pollution--some
can be connected only in a general way and others show specific
relationships. The effects are often classified as physical,
chemical, and biological.

Section II of this chapter treats the general effects. Particulates cause dirtiness, and they may contribute to climate
modification. Some claim that a new ice age will be caused by

particulate-reduced insolation, while others believe that carbon
dioxide will raise the temperature, melt the ice caps, and flood
the coastal areas. The chemical effects of air pollution are
most noted by the deterioration of exposed materials--corrosion
of metals, erosion or paints and surfaces, and the disintegration
of cloth materials such as nylon. There are biological effects
that may be correlated with air pollution by epidemiological
methods, directly or indirectly. The indirect correlations include
most respiratory diseases and symptoms and eye and nose irritation; for example, lung cancer is higher in urban than in rural
areas (by a factor of about 2), as are the usual air pollution
parameters, but an attempt to show that lung cancer was linked
with Los Angeles smog failed. The direct damage correlations
are much better defined in plants than in animals.

Section III of this chapter deals with the effects of specific
pollutants. The effects of particles on visibility have been studied and the ambient concentrations related to the visual range.
A 1-m^2 tube of air to the visual range normally contains about
1.8 gm of particles when the relative humidity is less than 70%.
The acute effects of many pollutants have been studied on test
animals and sometimes on people. The acute effects on animals
from ambient pollution are generally tenuous or nonexistent
because the normal ambient levels of air pollutants are below
those shown to cause physiological damage; however, levels
of ozone and oxidant approach the acute damage levels during
photochemical smog episodes. Noise and pollen are definitely
causing damage at the existing levels. The acute effects of a
number of pollutants on plants have been well established--
sulfur dioxide, fluorides, ethylene, ozone, and so on. Plants
are more sensitive to most pollutants than are animals.

There is a trend toward basing air quality criteria on levels
that cause physiological change (rather than damage), changes
such as Pavlovian responses and psychomotor function deteriora-

tions. This philosophy has been used in Russia for a long time
and is now being used to argue the effects of carbon monoxide
by several researchers in this country. Control of air pollution
is warranted because of esthetic insult and plant damage which
are quite evident, and the biological effects on animals which
are usually proved only by statistical epidemiological procedures
that are mistrusted by many should not be used to justify control
except in those instances in which effects are obvious, such as
damage to grazing cattle from fluorides from phosphate rock
operations or aluminum production or by arsenic from smelter
processes.

PROBLEMS

1. What is the usual composition of ambient air, neglecting
    components other than nitrogen, oxygen, argon, carbon
    dioxide, and water, when the relative humidity is 50%?

2. If Haber's law holds and 400 ppm_v causes a severe headache
    for a 2-hr exposure, how long an exposure to 1000 ppm_v
    will cause the same effect?

3. How many emphysema deaths occurred in the United States
    in 1962? How does this compare with 1952?

4. Document cases of Arndt-Schulz law from the literature.

5. How many cases of chest constriction from air pollution were
    reported by physicians in 1963?

6. Assume the effective particle diameter is 0.682 μm and its
    density is 2.5 gm/cm^3. Use Eqs. 10-10 and 10-13 to
    calculate the k value in Eq. 10-16.

7. If a state passes air pollution regulations which specify a
    minimum of 5 mi visibility, what should the suspended
    particulate limit be for consistency?

8. What is the surface area of 1 cm^3 of cubes that are 0.1 μm
    on an edge?

9. Show that 6 Cohs would represent an ambient concentration
    of about 2000 $μg/m^3$ suspended particulates.

10. What should the ambient carbon monoxide concentration be
    if 0.03 ppm_v equivalent is released per year and carbon
    monoxide has a half-life of 1 month? Assume steady-state
    conditions and equilibrium at 5 half-lives.

11. Derive Eq. 10-19 from Eq. 10-18.

12. If R is equal to 300 and 40% carboxyhemoglobin causes a certain physiological stress, what is the carbon monoxide concentration that causes the stress? How long must the exposure last?

13. Why would an ethylene producer put a cotton patch adjacent to his plant?

14. If the density of pollen grains is 1.2 gm/cm^3, how many 50-μm particles are contained in 1 lb?

15. What would be the temperature rise in the body caused by 600 rads (lethal dose) of ionizing radiation if the specific heat of the tissue is assumed to be the same as for water?

BIBLIOGRAPHY

"Air Pollution and Health," Am. Rev. Respiratory Diseases, 93:2, 12 pp., reprint, February 1966.

Air Quality Criteria for Particulate Matter, National Air Pollution Control Administration Publ. No. AP-49, Arlington, Virginia, January 1969.

Air Quality Criteria for Sulfur Oxides, National Air Pollution Control Administration Publ. No. AP-50, Arlington, Virginia, January 1969.

Air Quality Data--1964-1965, (NASN) U.S.P.H.S., Cincinnati, Ohio, 1966.

M. O. Amdur, "Toxicologic Appraisal of Particulate Matter, Oxides of Sulfur, and Sulfuric Acid," discussion by J. W. Clayton, Jr., J. Air Pollution Control Assoc., 19:9, 638-646, September 1969.

M. O. Amdur, "The Effect of Various Aerosols on the Response of Guinea Pigs to Sulfur Dioxide," Arch. Environmental Health, 16:4, 460-468, April 1968.

R. R. Beard, "Toxicological Appraisal of Carbon Monoxide," discussion by B. D. Dinman, J. Air Pollution Control Assoc., 19:9, 722-729, September 1969.

E. W. Burt, "A Study of the Relation of Visibility to Air Pollution," Am. Industrial Hyg. Assoc., 22:2, 102-108, March-April, 1961.

R. J. Charlson, "Atmospheric Visibility Related to Aerosol Mass Concentration--A Review," Env. Sci. Tech., 3:10, 913-918, October 1969.

Cleaning Our Environment--The Chemical Basis for Action, American Chemical Society, Washington, D. C., 1969.

"Community Air Quality Guides" appearing in Am. Industrial Hyg.
Assoc. J.: "Rationale," 29:1, 1-3, January-February 1968.
"Iron Oxide," 29:1, 4-6, January-February 1968. "Aldehydes," 29:5, 505-512, September-October 1968. "Ethylene," 29:6, 627-631, November-December 1968. "Lead,"
30:1, 95-97, January-February 1969. "Inorganic Fluorides,"
30:1, 98-101, January-February 1969. "Phenol and Cresol,"
30:4, 425-428, July-August 1969. "Particulates," 30:4,
428-434, July-August 1969. "Ozone," 29:3, 299-303, May-
June 1968. "Beryllium," 29:2, 189-192, March-April 1968.
"Carbon Monoxide," 30:3, 322-325, May-June 1969.

The Effects of Air Pollution, U.S.P.H.S. Publ. No. 1556,
Washington, D. C., 1966.

W. L. Faith, Air Pollution Control, Wiley, New York, 1959.

W. L. Faith, "Inert Particulates--Nuisance Effects," J. Occupational Med., 10:9, 539-544, September 1968.

M. Glasser and L. Greenburg, "Air Pollution is Linked Directly
to Death Rate," quoted by Gary Brooten, The Evening Bulle-
tin (Philadelphia), 4H, Tuesday, November 11, 1969.

J. R. Goldsmith and S. I. Cohen, "Epidemiological Bases for
Possible Air Criteria for Carbon Monoxide," J. Air Pollution
Control Assoc., 19:9, 704-713, September 1969.

G. M. Hidy and J. R. Brock, "Lung Deposition of Aerosols--A
Footnote on the Role of Diffusiophoresis," Environmental
Sci. Tech., 3:6, 563-567, June 1969.

L. S. Jaffe, "Photochemical Air Pollutants and Their Effects on
Men and Animals," Part I, Arch. Environmental Health,
15:12, 782-791, December 1967; Part II, 16:2, 241-255,
February 1968.

R. A. Kehoe, "Toxicological Appraisal of Lead in Relation to the
Tolerable Concentration in the Ambient Air," discussion by
H. L. Hardy, J. Air Pollution Control Assoc., 19:9, 690-
703, September 1969.

D. Kinsey and P. Bender, "Air Pollution and the Heart," J. Am.
Med. Women's Assoc., 21:7, 583-585, July 1966.

A. V. Kneese, "How Much is Air Pollution Costing Us in the
United States?" with discussion by B. Linsky, Proc. Third
National Conf. on Air Pollution, December 12-14, U.S.P.
H.S. Publ. No. 1649, Washington, D. C., 1966.

E. Landau, R. Smith, and D. A. Lynn, "Carbon Monoxide and
Lead--An Environmental Appraisal," discussion by G. D.
Clayton, J. Air Pollution Control Assoc., 19:9, 684-689,
September 1969.

P. A. Leighton, Photochemistry of Air Pollution, Academic Press, New York, 1961.

H. C. McKee, "Particulate Standards Keyed to Visibility and Citizen Complaints Are Interim Measure," discussion by F. W. Church, Environmental Sci. Tech., 3:6, 542, 546-548, June 1969.

W. Machle, "Major Challenge in Setting SO_2 Standards is Avoiding Simplistic Approach," discussion by H. N. MacFarland and R. E. Eckardt, Environmental Sci. Tech., 3:6, 543-545, June 1969.

P. L. Magill, F. R. Holden, and C. Ackley, Eds., Air Pollution Handbook: Section 6, C. Steffens, "Visibility and Air Pollution"; Section 7, J. J. Phair, "The Epidemiology of Air Pollution"; Section 9, M. D. Thomas and R. H. Hendricks, "Effect of Air Pollution on Plants," McGraw-Hill, New York, 1956.

P. K. Mueller and M. Hitchcock, "Air Quality Criteria--Toxicological Appraisal for Oxidants, Nitrogen Oxides, and Hydrocarbons," discussion by R. C. Wands, J. Air Pollution Control Assoc., 19:, 670-678, September 1969.

E. K. Peterson, "Carbon Dioxide Affects Global Ecology," Environmental Sci. Tech., 3:11, 1162-1169, November 1969.

J. M. Pierrard, "Environmental Appraisal--Particulate Matter, Oxides of Sulfur, and Sulfuric Acid," discussion by J. P. Lodge, Jr., and P. W. West, J. Air Pollution Control Assoc., 19:9, 658-669, September 1969.

F. Sargent, II, "Adaptive Strategy for Air Pollution," Bioscience, 17:10, 691-697, October 1967.

R. G. Smith, "Ambient CO Levels May Be No Immediate Health Threat--But What About Aesthetics?," discussion by B. Weinstock and G. M. Wilkening, Environmental Sci. Tech., 3:7, 628-638, July 1969.

F. E. Speizer, "An Epidemiological Appraisal of the Effects of Ambient Air on Health: Particulates and Oxides of Sulfur," discussion by I. J. Selikoff, J. Air Pollution Control Assoc., 19:9, 647-656, September 1969.

C. Starr, "Social Benefit versus Technological Risk," Science, 165, 1232-1238, September 19, 1969.

A. C. Stern, Ed., Air Pollution, Vol. I: Air Pollution and Its Effects, 2nd ed., Academic Press, New York, 1968.

J. Stewart-Gordon, "We're Poisoning Ourselves with Noise," Reader's Digest, 187-194, February 1970.

H. E. Stokinger, "The Spectre of Today's Environmental Pollution--USA Brand: New Perspectives from an Old Scout," Am. Industrial Hyg. Assoc. J., 30:3, 195-215, May-June 1969.

C. R. Thompson and O. C. Taylor, "Effects of Air Pollutants on Growth, Leaf Drop, and Yield of Citrus Trees," Environmental Sci. Tech., 3:10, 934-940, October 1969.

"Threshold Limit Values of Airborne Contaminants--1970," Am. Conf. of Governmental Industrial Hygienists, Cincinnati, Ohio, 1970.

Toxicologic and Epidemiologic Bases for Air Quality Criteria, Special Issue, J. Air Pollution Control Assoc., 19:9, September 1969.

J. B. Upham, "Materials Deterioration and Air Pollution," presented at 57th Air Pollution Control Association Meeting, Houston, June 1964.

M. I. Weisburd, "Physician's Guide to Air Pollution," J. Am. Med. Assoc., 186, 605-609, November 9, 1963.

J. D. Williams et al., Interstate Air Pollution Study: VI. Effects of Air Pollution, National Center for Air Pollution Control, Cincinnati, Ohio, December 1966.

G. W. Wright, "An Appraisal of Epidemiologic Data Concerning the Effect of Oxidants, Nitrogen Dioxide and Hydrocarbons upon Human Populations," discussion by R. M. Albrecht, J. Air Pollution Control Assoc., 19:9, 679-682, September 1969.

AUTHOR INDEX

Underlined numbers refer to Bibliography pages, where complete citations are given.

SUBJECT INDEX

A

B